Clinical Anatomy of the Cranial Nerves

Clinical Anatomy of the Cranial Nerves

Paul Rea

AMSTERDAM • BOSTON • HEIDELBERG • LONDON
NEW YORK • OXFORD • PARIS • SAN DIEGO
SAN FRANCISCO • SINGAPORE • SYDNEY • TOKYO

Academic Press is an imprint of Elsevier

Academic Press is an imprint of Elsevier
32 Jamestown Road, London NW1 7BY, UK
The Boulevard, Langford Lane, Kidlington, Oxford OX5 1GB, UK
Radarweg 29, PO Box 211, 1000 AE Amsterdam, The Netherlands
225 Wyman Street, Waltham, MA 02451, USA
525 B Street, Suite 1800, San Diego, CA 92101-4495, USA

First published 2014

Library of Congress Cataloging-in-Publication Data
A catalog record for this book is available from the Library of Congress

British Library Cataloguing-in-Publication Data
A catalogue record for this book is available from the British Library

ISBN: 978-0-12-800898-0

For information on all Elsevier publications
visit our website at **store.elsevier.com**

This book has been manufactured using Print On Demand technology. Each copy is produced to
order and is limited to black ink. The online version of this book will show color figures where
appropriate.

CONTENTS

Having personally studied Medicine and now in the privileged position of teaching human anatomy to medical, dental, and science students, one area that has always been a challenge to understand is the anatomy of the cranial nerves. They are perhaps the most complex set of nerves in the human body. Within such a small territory, they supply so many structures, some absolutely essential to life.

However, one thing that has been even more of a challenge was to take the anatomy of each nerve and understand the clinical application of that nerve or collection of nerves. Many resources are currently available for human anatomy training but not always in an appropriately succinct and clinically relevant fashion.

Therefore, this book is incredibly unique as a resource as each cranial nerve is dealt with on a chapter-by-chapter basis. The key anatomical points are highlighted including origins, relevant branches, divisions, ganglia, and relations in summary form for easier understanding and referencing. It then takes the reader through, in a step-by-step process, how to clinically examine each of the cranial nerves in turn after dealing with the underpinning anatomy. Useful hints and tips are provided from the clinical and academic expertise of the author to ensure the reader has an applied understanding, but also are able to communicate effectively to the patient during the examination. Following on from this, common pathologies of each of the cranial nerves are given to try and help explain clinical findings on examination. At the end of each chapter on a cranial nerve, interesting clinical questions and answers are dealt with to explain some of the reasoning behind the clinical presentation relating it back to the previous work, but also expanding ensuring the reader undertakes applied lateral thinking.

This book combines anatomical knowledge, pathology, clinical examination, and explanation of clinical findings, drawing together material which is in general scattered throughout anatomical textbooks, arising every time that a nerve is dealt with in other resources. This way, everything will be to hand for the learner in a format that will be used as a reference point, instructing on anatomy, but also how to examine the functioning of this anatomy in the patient.

It would be envisaged that this book would be a primary book as a quick reference guide, but it can also be used alongside the readers' other resources from their clinical experience and training, including lecture/tutorial notes, textbooks, and workbooks, as it would be pulling together material covered in so many fields including pathology, surgery, clinical examination, and anatomy.

For each of the chapters on the 12 pairs of the cranial nerves (i.e., 12 separate chapters for each nerve), the same format is used for each one:

- **The Anatomy—Summary**
  A brief summary statement giving the foundation knowledge of the nerve, its primary components, and what it supplies and its functions are.
- **The Anatomy—In More Detail**
  This section will detail the nerve's origin, course, intracranial pathway, site of exit from the skull, extracranial course, and to its termination. This section will give a succinct and comprehensive account of the important features of the nerve including its components, related ganglion (ganglia), important branches, and particular relations as relevant to the clinical applications related to that nerve. The purpose of this section is to provide the relevant anatomy that the reader will be able to apply to the patient at the bedside or other clinical setting. Tables will provide an easy way for the reader to identify key materials as relevant to each nerve. Images of digitally captured and professionally processed plastinated anatomical specimens will highlight the nerve, structure(s) it supplies, and anatomical relations to other structures.
- **The Clinical Application**
  This section is in two parts. The first section will be related to testing the nerve at the bedside. It is a simple step-by-step format providing an easy to read account of what to do. Included in this section are hints and tips as to best practice, based on the extensive experience of the author. This will help the examiner of the cranial nerves feel more confident and also instil an air of confidence and trust from the patient by proven hints and tips that are relevant and work. The second part of this section will be a brief comment on advanced testing which may involve more technical examinations

and equipment to be used on the patient or the relevant practitioner to contact for further specialist advice.

- **Pathologies**

  This section details the most common pathologies to affect each nerve. It will also describe (as relevant) the branches that can be affected at the site of a pathology, and how this will be manifest in the clinical presentation. Summary tables are provided for ease of finding material based on a patient's symptoms or nerve (branches) involved. This provides the reader with a form of mini-diagnosis, though should be seen as a guide subject to further investigation of the underlying pathology.

- **Interesting Clinical Questions**

  The final section will provide an opportunity for lateral thinking and consolidation of the knowledge gained from the section on each nerve. It will pose problems in a question and answer format making the reader think about the clinical applications of the anatomy and providing a detailed description of the reasoning from an anatomical and clinical perspective.

OVERALL PURPOSE

This book is intended to enhance learning and enable the learner to take control of understanding the cranial nerves not just from the anatomical perspective but also the applied clinical examination and related pathologies which may be encountered. It can be used by a wide variety of specialists who are involved in detailed examination of a patient's cranial nerves. I hope it presents this information in a readily accessible, easy-to-understand format to inspire the next generations' understanding of the truly amazing cranial nerves.

ACKNOWLEDGMENTS

There are several people whom I would like to thank in making this book possible.

First I would like to express my gratitude to Elsevier for having the time, patience, and faith in me while putting this together. They have really been the backbone to helping me realize my dream in publishing this.

I would like to dedicate this book to my mother Nancy, father Paul, and dearest brother Jaimie. Thank you for being there and supporting me throughout everything—I am so proud of you all! Thank you also to Jennifer Rea—my sister-in-law—but more like a sister to me.

I would also like to extend a special note of thanks to David Kennedy. Also, thanks to Christine and Susan Kennedy!

Thank you to a dear friend who has gone but not been forgotten— Mark Peters. I am also most grateful to all the years of great friendship and banter that Dr. Richard Locke has given me. Cheers!

A note of enormous thanks goes to Ms. Caroline Morris who helped to create the images throughout this book. Without her help it would not have been possible to create these high-quality pictures.

Also, thank you to a dear colleague and amazing mentor who has supported me from when I first started my career as an anatomist through to where I am today—Dr. John Shaw-Dunn.

Finally, thank you to our donors and their families.

OVERVIEW OF THE NERVOUS SYSTEM

Broadly speaking, the nervous system is divided into two components—central and peripheral. The central nervous system (CNS) comprises the brain and the spinal cord. The peripheral nervous system (PNS) comprises all of the nerves—cranial, spinal, and peripheral nerves, including the sensory and motor nerve endings.

DIVISIONS OF THE NERVOUS SYSTEM

Central Nervous System

The CNS is comprised of the brain and spinal cord. The purpose of the CNS is to integrate all the body functions from the information it receives. Within the PNS, there are abundant nerves (group of many nerve fibers together); however, the CNS does not contain nerves. Within the CNS, a group of nerve fibers traveling together is called a pathway or tract. If it links the left- and right-hand sides it is referred to as a commissure.

Neurons

Within the CNS, there are many, many millions of nerve cells called neurons. Neurons are cells which are electrically excitable and transmit information from one neuron to another by chemical and electrical signals. There are three very broad classifications of neurons—sensory (which process information on light, touch, and sound to name some sensory modalities), motor (supplying muscles), and interneurons (which interconnect neurons via a network).

Typically, a neuron comprises a few basic features. In general, a neuron has a cell body. Here, the nucleus—the powerhouse—of the neuron lies with its cytoplasm. At this point, numerous fine fibers enter called dendrites. These processes receive information from adjacent neurons keeping it up-to-date with the surrounding environment. This way the amount of information that a single neuron receives is significantly increased. From a neuron, there is a long single process of variable length called an axon. This conducts information away from the neuron. Some neurons however have no axons and the dendrites will conduct information to and from the neuron. In addition to this, a lipoprotein layer called the *myelin sheath* can surround the axon of a principal cell. This is not a continuous layer along the full length of the axon. Rather, there are interruptions called *nodes of Ranvier*. It is at this point where the voltage-gated channels occur, and it is at that point where conduction occurs. Therefore, the purpose of the myelin sheath is to enable almost immediate conduction between one node of Ranvier and the next ensuring quick communication between neurons.

In relation to the size of neurons, this varies considerably. The smallest of neurons can be as small as 5 μm, and the largest, for example, motor neurons, can be as big as 135 μm. In addition, axonal length can vary considerably too. The shortest of these can be 100 μm, whereas a motor axon supplying the lower limb, for example, the toes, can be as long as 1 m.

In the PNS, neurons are found in *ganglia*, or in *laminae* (layers) or *nuclei* in the CNS.

Neurons communicate with each other at a point called a synapse. Most of these junctional points are chemical synapses where there is the release of a neurotransmitter which diffuses across the space between the two neurons. The other type of synapse is called an electrical synapse. This form is generally more common in the invertebrates, where there is close apposition of one cell membrane and the next (i.e., at the pre- and postsynaptic membranes). Linking these two cells is a collection of tubules called *connexons*. The transmission of impulses occurs in both directions and very quickly. This is because there is no delay in the neurotransmitter having to be activated and released across the synapse. Instead, the flow of communication depends on the membrane potentials of the adjacent cells.

Neuroglia
Neuroglia, or glia, are the supportive cells for neurons. Their main purpose is not in relation to the transmission of nerve impulses. Rather, they are involved in providing nutrient support, maintenance of homeostasis, and the production of the myelin sheath. There are two broad classifications—microglia and macroglia.

The microglia have a defense role as a phagocytic cell. They are found throughout the brain and spinal cord and can change their shape, especially when they engulf particulate material. They are therefore serving a protective role for the nervous system.

Macroglia are subdivided into seven different types, again with each having a special role.

1. *Astrocytes*
   These cells fill in the spaces between neurons and provide for structural integrity. They also have processes which join to the capillary blood vessels. These are known as *perivascular end feet*. Therefore, with their close apposition to blood vessels, they are also thought to be responsible for metabolite exchange between the neurons and the vasculature. They are found in the CNS.
2. *Ependymal cells*
   There are three types of ependymal cells—ependymocytes, tanycytes, and choroidal epithelial cells. The ependymocytes allow for the free movement of molecules between the cerebrospinal fluid (CSF) and the neurons. Tanycytes are generally found in the third ventricle and can be involved in responding to changing hormonal levels of the blood-derived hormones in the CSF. Choroidal epithelial cells are the cells which control the chemical composition of the CSF. They are found in the CNS.
3. *Oligodendrocytes*
   These cells are responsible for the production of myelin sheaths. They are found in the CNS.
4. *Schwann cells*
   Like oligodendrocytes, Schwann cells are responsible for the production of the myelin sheath, but in the PNS. They also have an additional role in phagocytosis of any debris, therefore help to clean the surrounding environment.
5. *Satellite cells*
   These cells surround those neurons of the autonomic system and also the sensory system. They maintain a stable chemical balance of the surrounding environment to the neurons. They are therefore found in the PNS.
6. *Radial glial cells*
   Radial glial cells act as scaffolding onto which new neurons migrate to. They are found in the CNS.

7. *Enteric glia cells*

These cells are found within the gastrointestinal tract and aid digestion and maintenance of homeostasis. They are by their very nature found in the PNS.

Grey and White Matter

In the CNS, there are two clear differences between the structural components. It is divided into its appearance of either grey or white matter. Within the grey matter, there are cell bodies and dendrites of efferent neurons, glial cells (supportive), fibers of afferent neurons, and interneurons. The white matter on the other hand primarily consists of myelinated axons and the supportive glial cells. The purpose of the white matter is to allow for communication from one part of the cerebrum to the other and also to communicate to other brain areas and carry impulses through the spinal cord.

Brain

The brain is comprised of three swellings which form during development—the forebrain (prosencephalon), midbrain (mesencephalon), and hindbrain (rhombencephalon). During development in mammals, the forebrain continues to grow, whereas in other vertebrates, for example, amphibians and fish, the three divisions remain in proportion to each other during growth.

The brain can also be subdivided into the following:

a. *Telencephalon* (cerebral hemispheres) + *Diencephalon* (thalamus and hypothalamus) = FOREBRAIN
b. *Mesencephalon* = MIDBRAIN
c. *Metencephalon* (pons, cerebellum, and the trigeminal, abducens, facial, and vestibulocochlear nerves) + *Myelencephalon* (medulla oblongata) = HINDBRAIN

Forebrain Functions

Surrounding the core of the forebrain, that is, the diencephalon, are the two large cerebral hemispheres (left and right), which constitute the cerebrum. The cerebrum is comprised of three regions:

1. Cerebral cortex

The cerebral cortex is the grey matter of the cerebrum. It is comprised of three parts based on its functions—motor, sensory, and association areas. The motor area is present in both cerebral cortices. Each one controls the opposite side of the body, that is, the left motor area controls the right side of the body, and vice versa. There are two broad regions—a primary motor area responsible for execution of voluntary movements and supplementary areas involved in selection of voluntary movements.

The sensory area receives information from the opposite side of the body, that is, the right cerebral cortex receives sensory information from the left side of the body. In essence it deals with auditory information (via the primary auditory cortex), visual information (via the primary visual cortex), and sensory information (via the primary somatosensory cortex).

The association areas allow us to understand the external environment. All of the cerebral cortex is subdivided into lobes of the brain. These are as follows:

a. Frontal lobes

Broadly speaking the frontal lobe deals with "executive" functions and our long-term memory. It also is the site of our primary motor cortex, toward its posterior part.

b. Parietal lobes

The parietal lobes are responsible for integration of sensory functions. It is the site of our primary somatosensory cortex.

c. Temporal lobes

The temporal lobes integrate information related to hearing; and therefore, it is the site of our primary auditory cortex.

d. Occipital lobes

The occipital lobes integrate our visual information and functions as the primary visual cortex.

2. Basal ganglia

The basal ganglia are three sets of nuclei—the *globus pallidus*, *striatum*, and *subthalamic nucleus*. These nuclei are found at the lower end of the forebrain and are responsible for voluntary movement, development of our habits, eye movements, and our emotional and cognitive functions.

3. Limbic system

The limbic system is comprised of a variety of structures on either side of the thalamus. It serves a variety of functions including long-term memory, processing of the special sense of smell (olfaction), behavior, and our emotions.

Thalamus

The thalamus is like a junction point of information. It is a relay point for all sensory information (apart from that related to smell). It also functions in the regulation of our wakened state or sleep. In addition, it provides a connection point for motor information on its way to the cerebellum.

Hypothalamus

The hypothalamus, as its name suggests, is located below the thalamus. It secretes hormones influencing the pituitary gland, and in turn, a wide variety of bodily functions. It regulates autonomic activity ranging from temperature control, hunger, and our circadian rhythm and thirst.

Midbrain Functions

The midbrain, as its name suggests, is found between the hindbrain below and the cerebral cortices above. Comprised of the *cerebral peduncles*, *cerebral aqueduct*, and the *tegmentum*, it is involved in motor function, arousal state, temperature control, and visual and hearing pathways.

Hindbrain Functions

The lowest part of the brain developmentally is the hindbrain and comprises the pons, medulla, and the cerebellum. These areas control movement, cardiorespiratory functions, and a variety of bodily functions like hearing and balance, facial movement, swallowing, and bladder control. Therefore, brainstem death, that is, death of these regions, is incompatible with life.

Spinal Cord

The grey matter is surrounded by the white matter which contains primarily the axons of myelinated interneurons. These groups of axons, or pathways, run longitudinally either up and down to and from the brain or between upper and lower levels of the spinal cord.

Surrounding the spinal cord are three layers of protective meninges—pia mater, arachnoid mater, and dura mater. The dural layer is the outermost layer and extends from approximately the second sacral segment all the way to the foramen magnum, and into the brain to also surround that too. The arachnoid layer is deeper and is tightly adherent to the dura mater. It therefore leaves a space just deep to it called the subarachnoid space. It is in this space that the fluid that circulates around the brain and spinal cord passes to cushion, support, protect, and provide nourishment to the brain and spinal cord—the CSF. The innermost layer is called the pia mater and is tightly adherent to the spinal cord and also to the brain.

The spinal cord extends from the brain to the second lumbar segment, but the spinal nerves still emerge from the corresponding vertebrae all the way to the coccyx. Toward the lower end of the spinal cord, there is an enlargement called the lumbosacral segment and this constitutes the *cauda equina* (Latin for horse's tail) is that part of the lower spinal cord from the second to fifth lumbar vertebral nerves. The spinal cord terminates as the *filum terminale*. This relationship of where the spinal cord terminates has a clinical relevance—*lumbar puncture*.

As the core of the spinal cord terminates at approximately the level of the second vertebral level, the clinician is able to insert a needle below this point to take a sample of CSF. At the level of the third and fourth vertebral level space, a needle can be inserted into the subarachnoid space to take a sample of CSF. This can be taken for analysis to diagnose, or otherwise, a potential infection of the meninges—*meningitis*, which can be potentially life threatening.

When the spinal cord is examined in transverse section, it is comprised of a central grey matter (butterfly shaped) comprising cell columns oriented along the rostro-caudal axis (containing neuronal cell bodies, dendrites, and axons that are both myelinated and unmyelinated), surrounded by the white matter comprising the ascending and descending myelinated and unmyelinated fasciculi (tracts). The general layout of the spinal cord is shown in Figure I.1.

In each half of the spinal cord, there are three funiculi: the dorsal funiculus (between the dorsal horn and the dorsal median septum), the lateral funiculus (located where the dorsal roots enter and the ventral roots exit), and the ventral funiculus (found between the ventral median fissure and the exit point of the ventral roots).

Based on detailed studies of neuronal soma size (revealed using the Nissl stain), Rexed (1952) proposed that the spinal grey matter is arranged in the dorsoventral axis into laminae and designated them into 10 groupings of neurons identified as I–X.

Lamina I contains the terminals of fine myelinated and unmyelinated dorsal root fibers that pass first through the zone of Lissauer (dorsolateral funiculus) and then enter lamina I mediating pain and temperature sensation (Christensen and Perl, 1970; Menétrey et al., 1977; Craig and Kniffki, 1985; Bester et al., 2000). The neurons here

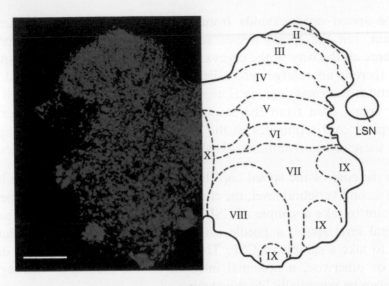

Figure I.1 The general layout of the spinal cord. This section is taken at the fifth lumbar vertebral level using immunofluorescence on the left side for a neuronal marker NeuN. The neuronal cells are highlighted in red within the spinal cord section. The diagrammatic representation on the right shows Rexed's laminae (1952) numbered I–X. A small nucleus is also found out in the dorsolateral white matter called the lateral spinal nucleus (LSN). The scale bar represents 500 μm.

have been divided into small neurons and large marginal cells characterized by wide-ranging horizontal dendrites (Willis and Coggeshall, 1991). They then synapse on the posteromarginal nucleus. From here the axons of these cells pass to the opposite side and ascend as the lateral spinothalamic tract.

Lamina II is immediately below lamina I, referred to as the substantia gelatinosa. Neurons here modulate the activity of pain and temperature afferent fibers, though intrinsic neurons here do not contain the target for substance P, the NK-1 receptor (Bleazard et al., 1994), which is however found in lamina I, III, and IV (Naim et al., 1997). Lamina II has been subdivided into an outer (dorsal) lamina II (II_o) and an inner (ventral) lamina II (II_i) based on the morphology of these layers with stalked cells found in larger numbers in lamina II_o but stalked and islet cells were found throughout lamina II_i (Todd and Lewis, 1986). Indeed, lamina II_i was also found to be different in its neurochemical profile. Lamina II is the region which receives an extensive unmyelinated primary afferent input, with very little from large myelinated primary afferents (except for distal parts of hair follicle afferents in some animals; Willis and Coggeshall, 1991). The axonal projections from here are wide and varied with some neurons project-ing from the spinal cord (projection neurons), some passing to different laminae and some with axons confined to a lamina in the region of the dendritic tree of that cell (e.g., intralaminar interneurons, local inter-neurons, and Golgi type II cells) (Todd, 1996).

Lamina III is distinguished from lamina II in that it has slightly larger cells, but with a neuropil similar to that of lamina II. The classi-cal input to this lamina comes from hair follicles and other types of coarse primary afferent fibers which include Pacinian corpuscles and rapidly and slowly adapted fibers.

Lamina IV is a relatively thick layer that extends across the dorsal horn. Its medial border is the white matter of the dorsal column, and its lateral border is the ventral bend of laminae I—III. The neurons in this layer are of various sizes ranging from small to large and the affer-ent input here is from collaterals and from large primary afferent fibers (Willis and Coggeshall, 1991). Input also arises from the substantia gelatinosa (lamina II) and contributes to pain, temperature, and crude touch via the spinothalamic tract (Siegel and Sapru, 2006).

Lamina V extends as a thick band across the narrowest part of the dorsal horn. It occupies the zone often called the neck of the dorsal horn. It has a well demarcated edge against the dorsal funiculus, but an indistinct lateral boundary against the white matter due to the many longitudinally oriented myelinated fibers coursing through this area. The cell types are very homogeneous in this area, with some being slightly larger than in lamina IV (Willis and Coggeshall, 1991). Again, like lamina IV, primary afferent input into this region is from large primary afferent collaterals as well as receiving descending fibers from the corticospinal and rubrospinal tracts with axons also contrib-uting to the spinothalamic tracts (Siegel and Sapru, 2006). In addition, in the thoracolumbar segments (T1—L2/3) the reticulated division of lamina V contains projections to sympathetic preganglionic neurons (Cabot et al., 1994).

Lamina VI is present only in the cervical and lumbar segments. Its medial segment receives joint and muscle spindle afferents, with the lateral segment receiving the rubrospinal and corticospinal pathways. The neurons here are involved in the integration of somatic motor processes.

Lamina VII present in the intermediate region of the spinal grey matter contains Clarke's nucleus extending from C8–L2. This nucleus receives tendon and muscle afferents with the axons of Clarke's nucleus forming the dorsal spinocerebellar tract relaying information to the ipsilateral cerebellum (Snyder et al., 1978). Also within lamina VII are the sympathetic preganglionic neurons constituting the intermediolateral cell column in the thoracolumbar (T1–L2/3) and the parasympathetic neurons located in the lateral aspect of the sacral cord (S2–4). In addition, Renshaw cells are located in lamina VII and are inhibitory interneurons which synapse on the alpha motor neurons and receive excitatory collaterals from the same neurons (Renshaw, 1946; Siegel and Sapru, 2006).

Lamina VIII and IX are found in the ventral grey matter of the spinal cord. Neurons here receive descending motor tracts from the cerebral cortex and the brainstem and have both alpha and gamma motor neurons here which innervate skeletal muscles (Afifi and Bergman, 2005). Somatotopic organization is present where those neurons innervating the extensor muscles are ventral to those innervating the flexors, and neurons innervating the axial musculature are medial to those innervating muscles in the distal extremities (Siegel and Sapru, 2006).

Lamina X is the grey matter surrounding the central canal and represents an important region for the convergence of somatic and visceral primary afferent input conveying nociceptive and mechanoreceptive information (Nahin et al., 1983; Honda, 1985; Honda and Perl, 1985). In addition, lamina X in the lumbar region also contains preganglionic autonomic neurons as well as an important spinothalamic pathway (Ju et al., 1987a,b; Nicholas et al., 1999).

Peripheral Nervous System

The nerves in the PNS transmit information from all parts of the body to and from the CNS. In total, there are 43 pairs of nerves in the PNS—12 cranial nerves and 31 spinal nerves.

The nerves of the PNS can be either myelinated (formed by the surrounding Schwann cells) or unmyelinated in nature. Whether they have this myelin or not, they do have the same general feature in that a nerve contains nerve fibers with axons of either afferent or efferent neurons. Therefore, the nerves of the PNS can be classified as belonging to either afferent (taking information to the CNS) or efferent (away from the CNS). With spinal nerves, they contain both afferent and efferent information, whereas some cranial nerves like the olfactory and optic nerves contain only afferent information (for smell and sight, respectively).

Afferent information transmits impulses from receptors to the CNS. Their axons are found outwith the CNS, but then enter the CNS. Efferent information however transmits information from the CNS externally to, for example, glands and muscles. It is worth noting that the efferent division is subclassified into what they ultimately supply. The further classification used of the efferent division is the somatic and autonomic nervous system (ANS). Simply put, the somatic nervous system innervates skeletal muscle, whereas the ANS innervates glands, neurons of the gastrointestinal tract, and cardiac and smooth muscles of glandular tissue.

Somatic Nervous System

The somatic nervous system consists of the cell bodies located in either the brainstem or the spinal cord. They have an extremely long course as they do not synapse after they leave the CNS until they are at their termination in skeletal muscle. They consist of large diameter fibers and are ensheathed with myelin. They are commonly referred to as motor neurons due to their termination in skeletal muscle. Within the muscle fibers, they release the neurotransmitter acetylcholine and are only excitatory, that is, result only in contraction of the muscle.

Cranial Nerves

There are 12 pairs of cranial nerves and these emerge from either the brain as fiber tracts (olfactory (I) and optic (II) nerves) or the brainstem (all other cranial nerves (III–XII)). The details of each of these nerves, their functions, and clinical applications will be dealt with in turn in subsequent chapters.

Spinal Nerves

There are 31 pairs of nerves that connect with the spinal cord as the spinal nerves. There are 8 in the cervical region, 12 in the thoracic region, 5 in the lumbar region, 5 in the sacral region, and 1 coccygeal nerve.

Autonomic Nervous System

The ANS can be thought of as that part of the nervous system supplying all other structures apart from skeletal muscle (supplied by the somatic nervous system). However, part of the ANS supplies the gastrointestinal system and is referred to as the enteric nervous system as the neurons are found supplying the glands and smooth muscle in the actual wall of the tract.

Within the ANS generally, it is comprised of two neurons and a synapse. This is different to the single neuron of the somatic nervous system. The origin of the first neuron of the ANS is found in the CNS, with the first synapse occurring in an autonomic ganglion. This part is defined as the preganglionic fiber. After the synapse in the autonomic ganglion, the second fiber is referred to as the postganglionic fiber as it passes to the effector organ, in this case cardiac or smooth muscle, glands or gastrointestinal neurons.

The ANS is subdivided into sympathetic and parasympathetic divisions based on physiological and anatomical differences. The sympathetic division arises from the thoracolumbar region from the first thoracic to the second lumbar level (T1−L2). The parasympathetic division arises from cranial and sacral origins. Specifically, the parasympathetic division arises from four of the cranial nerves—the oculomotor (III), facial (VII), glossopharyngeal (IX), and vagus (X) nerves. It also arises from the sacral plexus at the levels of the second to fourth sacral segments (S2−4).

Sympathetic Nervous System
The sympathetic nervous system arises from the thoracolumbar region of the spinal cord. Most of the sympathetic ganglia lie in close proximity to the spinal cord forming two chains on either side of the body. These are referred to as the *sympathetic trunks*. However, some ganglia lie a little further away from the spinal cord and are referred to as *collateral ganglia*. These are found close to the arteries in the abdomen with the same names (i.e., celiac, superior mesenteric, and inferior mesenteric ganglia). They tend to lie closer to the organs that they supply.

Although the sympathetic nervous system arises specifically at the thoracolumbar region, that is, from the first thoracic to either the second or third lumbar vertebral levels, the sympathetic trunk extends from the neck to the sacrum. This is because some of the preganglionic fibers arising from the thoracolumbar region travel either up or down several vertebral segments before forming their synapses with the respective postganglionic neurons. Within the neck, the cervical ganglia are referred to as the superior and middle cervical or stellate ganglia.

This allows the sympathetic nervous system to act as a single unit, but with small areas also able to act independently. This contrasts to the parasympathetic nervous system which tends to act independently. This arrangement is ideal to be involved in fine regulation of the activities of the organs or territories that they supply.

The sympathetic nervous system is responsible for the body's "fight or flight" reaction. Therefore, through its innervation of the adrenal medulla, which releases adrenaline (epinephrine) as its major secretion (80%; with the other 20% being noradrenaline (norepinephrine)), it helps to protect the body in times of threat to it. It would therefore be involved in functions like dilating the pupil, increasing heart rate and contractility, relaxation of the bronchial muscle and reduction in secretion of the bronchial glands, and reduction of gut motility. This allows blood to be diverted to those areas if the body needs to "fight or flight." The adrenal medulla is a bit unusual in its innervation by the sympathetic nervous system, as the postganglionic side of the adrenal medulla never develops axons. Instead the preganglionic fibers terminating in the adrenal medulla result in the secretion of epinephrine/norepinephrine, and is viewed as an endocrine gland as its secretions pass into the bloodstream.

Parasympathetic Nervous System
The parasympathetic nervous system is described as originating in the craniosacral region, that is, from the brainstem and also the sacral plexus. Specifically, the parasympathetic nervous system cranially is concerned with three of the cranial nerves, which will be dealt with in turn in greater detail throughout this book. The cranial nerves involved in the parasympathetic nervous system are the oculomotor, facial, glossopharyngeal, and vagus nerves. Specifically, the nuclei related to these are the Edinger–Westphal nucleus for the oculomotor nerve, superior salivatory and lacrimal nuclei for the facial nerve, inferior salivatory nucleus for the glossopharyngeal nerve, and the dorsal nucleus of the vagus nerve, as well as the nucleus ambiguus for the vagus nerve. This is where the preganglionic fibers are found for the parasympathetic nervous system. In addition to this, the sacral parasympathetic nucleus arising from the second, third, and fourth sacral segments is also involved.

The parasympathetic nervous system is opposite in its functions generally to the sympathetic nervous system. It can informally be referred to as the part of the nervous system responsible for "rest and digest," that is, responsible for the internal functions when you are sitting resting and relaxing. Therefore, it would constrict the pupil, slow heart rate and contractility, contract bronchial musculature and stimulate bronchial secretions, and enhance gut motility for digestion to effectively occur.

The main neurotransmitter in both the sympathetic and parasympathetic nervous systems at the preganglionic fiber, as it contacts the postganglionic fiber, is acetylcholine. The same is also true at the postganglionic fiber as it contacts the effector organ generally. Therefore, where acetylcholine is secreted, it is referred to as *cholinergic*. However, in the sympathetic nervous system, the major neurotransmitter between the postganglionic fiber and the effector organ tends to be noradrenaline (norepinephrine). It also tends to be the case that this is not an exclusive relationship as to what is secreted, and at what site it is secreted. In addition to this, cotransmitters tend to also be present (e.g., ATP, dopamine, and other neuropeptides).

REFERENCES

Afifi, A., Bergman, R., 2005. Functional Neuroanatomy, second ed. McGraw-Hill, New York, NY 10011.

Bester, H., Chapman, V., Besson, J.M., Bernard, J.F., 2000. Physiological properties of the lamina I spinoparabrachial neurons in the rat. J. Neurophysiol. 83, 2239–2259.

Bleazard, L., Hill, R.G., Morris, R., 1994. The correlation between the distribution of the NK-1 receptor and the action of tachykinin agonists in the dorsal horn of the rat indicates that substance P does not have a functional role on substantia gelatinosa (lamina II) neurons. J. Neurosci. 14 (12), 7655–7664.

Cabot, J.B., Alessi, V., Carroll, J., Ligorio, M., 1994. Spinal cord lamina V and lamina VII interneuronal projections to sympathetic preganglionic neurons. J. Comp. Neurol. 347, 515–530.

Christensen, B.N., Perl, E.R., 1970. Spinal neurons specifically excited by noxious or thermal stimuli: marginal zone of the dorsal horn. J. Neurophysiol. 33, 293–307.

Craig, A.D., Kniffki, K.D., 1985. Spinothalamic lumbosacral lamina I cells responsive to skin and muscle stimulation in the cat. J. Physiol. 365, 197–221.

Honda, C.N., 1985. Visceral and somatic efferent convergence onto neurons near the central canal in the sacral spinal cord of the cat. J. Neurophysiol. 53, 1059–1078.

Honda, C.N., Perl, E.R., 1985. Functional and morphological features of neurons in the midline region of the caudal spinal cord of the cat. Brain Res. 340, 285–295.

Ju, G., Hökfelt, T., Brodin, E., Fahrenkrug, J., Fischer, J.A., Frey, P., et al., 1987a. Primary sensory neurons of the rat showing calcitonin gene-related peptide immunoreactivity and their relation to substance P, somatostatin, galanin, vasoactive intestinal polypeptide and cholecystokinin immunoreactive ganglion cells. Cell Tissue Res. 247, 417–431.

Ju, G., Melander, T., Ceccatelli, S., Hökfelt, T., Frey, P., 1987b. Immunohistochemical evidence for a spinothalamic pathway co-containing cholecystokinin and galanin like immunoreactivities in the rat. Neuroscience 20, 439–456.

Menétrey, D., Giesler Jr., G.J., Besson, J.M., 1977. An analysis of response properties of spinal dorsal horn neurons to non-noxious and noxious stimuli in the rat. Exp. Brain Res. 27, 15–33.

Nahin, R.L., Madsen, A.M., Giesler, G.J., 1983. Anatomical and physiological studies of the grey matter surrounding the spinal cord central canal. J. Comp. Neurol. 220, 321–335.

Naim, M., Spike, R.C., Watt, C., Shehab, A.S., Todd, A.J., 1997. Cells in laminae III and IV of the rat spinal cord that possess the neurokinin-1 receptor and have dorsally directed dendrites receive a major synaptic input from tachykinin containing primary afferents. J. Neurosci. 17 (14), 5536–5548.

Nicholas, A.P., Zhang, X., Hökfelt, T., 1999. An immunohistochemical investigation of the opioid cell column in lamina X of the male rat lumbosacral spinal cord. Neurosci. Lett. 270, 9–12.

Renshaw, B., 1946. Central effects of centripetal impulses in axons of spinal ventral roots. J. Neurophysiol. 9, 191–204.

Rexed, B., 1952. The cytoarchitectonic organisation of the spinal cord in the cat. J. Comp. Neurol. 96, 414–495.

Siegel, A., Sapru, H.N., 2006. Essential Neuroscience. Lippincott Williams & Wilkins, Baltimore, MD.

Snyder, R.L., Faull, R.L., Mehler, W.R., 1978. A comparative study of the neurons of origin of the spinocerebellar afferents in the rat, cat and squirrel monkey based on the retrograde transport of horseradish peroxidase. J. Comp. Neurol. 15, 833–852.

Todd, A.J., 1996. GABA and glycine in synaptic glomeruli of the rat spinal dorsal horn. Eur. J. Neurosci. 8, 2492–2498.

Todd, A.J., Lewis, S.G., 1986. The morphology of Golgi-stained neurons in lamina II of the rat spinal cord. J. Anat. 149, 113–119.

Willis, W.D., Coggeshall, R.E., 1991. Sensory Mechanisms of the Spinal Cord, second ed. Plenum Press, New York, NY.

CHAPTER 1

Olfactory Nerve

THE ANATOMY—SUMMARY

The first cranial nerve is the olfactory nerve. It is the shortest of all the cranial nerves and is one of the two nerves that do not join with the brainstem. It is the cranial nerve responsible for conduction of impulses related to the special sense of smell (i.e., special sensory). Specifically, it is a special visceral afferent nerve.

THE ANATOMY—IN MORE DETAIL

The cell bodies of the olfactory receptor neurons are located in the olfactory organ. These are found in the upper part of the nasal cavity, nasal septum, and on the inner aspect of the superior nasal concha. The olfactory receptor nerves have two surfaces: basal and apical, and these receive information from the odors which then dissolve in the mucous fluid to allow for electrical transmission of those impulses.

When the odor dissolves in the mucous, it is then detected by the olfactory nerves. Each olfactory nerve has two components: an apical and a basal division. It is the cilia on the apical portion that detects the dissolved "smell." This then passes to the basal portion which constitutes the main processes of the olfactory nerve. The olfactory nerves at that point then enter the cranial cavity via the cribriform plate of the ethmoid bone (Figure 1.1). The special sense of smell is then transmitted toward the olfactory bulb where the cells then synapse. The transmission of the impulses carries posteriorly toward the brain via the olfactory tract. Specifically, the information passes to the piriform cortex of the anterior temporal lobe, anterior olfactory nucleus, amygdala, and entorhinal cortex.

The Components

The olfactory nerve is comprised of a single component allowing it to perform its role—special sensation (i.e., smell). The following diagram summarizes the pathway of innervation.

Clinical Anatomy of the Cranial Nerves. DOI: http://dx.doi.org/10.1016/B978-0-12-800898-0.00001-4

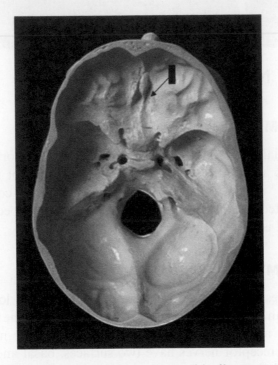

Figure 1.1 The dark block and the arrow point to the point of entry of the olfactory nerve through the cribriform plate of the ethmoid bone.

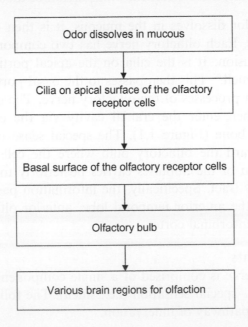

Odor dissolves in mucous

↓

Cilia on apical surface of the olfactory receptor cells

↓

Basal surface of the olfactory receptor cells

↓

Olfactory bulb

↓

Various brain regions for olfaction

Special Sensory
- Smell from the nasal septum, superior concha, and the roof of the nasal cavity. It is a special visceral afferent nerve.

The Important Branches
Within the nasal cavity, there are two types of fibers—those of the trigeminal nerve (see Chapter 5) which responds to irritating substances and temperature and those of the olfactory nerve for olfaction.

The olfactory receptor neurons have two parts:

1. Olfactory neurons in the olfactory epithelium have central processes which pass to the olfactory bulb where they synapse.
2. The processes from the synapse pass via the olfactory tracts to the brain regions for olfaction.

THE CLINICAL APPLICATION

Tip!

Always ensure you take a detailed history from the patient prior to doing any examinations to provide you with a background story and to help direct you to the most appropriate examination(s). ALWAYS introduce yourself to the patient, stating your name and position.

If undertaking clinical examination, give clear, simple, concise instructions if you need the patient to do anything to help elicit signs and/or symptoms.

Testing of the olfactory nerve is often missed in routine clinical examination. As it is involved in the special sensation of smell, testing of this nerve is undertaken by using a substance with a recognizable smell.

Tip!

It is handy to first evaluate the patency of the nasal passages bilaterally by asking the patient to breathe in through their nose while the examiner occludes one nostril at a time. Always tell the patient what you are going to do before doing it!

Testing at the Bedside

- Tell the patient that you want to test their sense of smell and ask permission to do so. Use vanilla essence, coffee, orange peel, or lemon juice.
- Then do the following:
  1. Ask them to cover one nostril at a time and then close their eyes and present the testing substance to each nostril. The testing substance must not be visible to the patient.
  2. Ask the patient to report if they smell anything. This allows identification of the ability to detect an odor. Asking them to identify what it is involves an "olfactory memory," that is, higher cortical functioning, only if they recognize the substance. DO NOT touch the patient when doing so and place it within 15–30 cm from the nasal cavity.
  3. Do the same for the opposite nostril.

●●●
___

Tip!

DO NOT use any odors that are irritating (e.g., menthol, ammonia) as these substances can stimulate the trigeminal nerve as well as the olfactory nerve resulting in a false-positive response.

___

PATHOLOGIES

There are three broad classifications of pathologies that can affect the olfactory nerve and the sense of smell: overactive (hyperosmia), underactive (hypoosmia), dysfunction (dysosmia), and absent (anosmia). Out of these categories, the most common is hypoosmia due because, for example, the common cold leading to rhinitis. The following table provides a few common pathologies that can result in a change of smell for the patient.

Classification	Common Pathologies
Hypoosmia	Common cold/allergy Smoking Nasal polyps Diabetes mellitus
Hyperosmia	Difficult to ascertain from a patient. Possible causes could be genetic or occasionally in patients with Addison's disease
Anosmia	Can be unilateral or bilateral: Head trauma (most common) Brain tumors (especially of the anterior cranial fossa) Cerebral infections (e.g., meningitis)
Dysosmia	Neurological disorders (e.g., Parkinson's, Alzheimer's disease, and Huntington's chorea)

INTERESTING CLINICAL QUESTIONS

Q:

What clinical conditions can result in dysfunction of the olfactory nerve?

A:

The simple answer to this is many! Below is a diagram to summarize some conditions which can result in changed olfactory nerve function, dependent on what body system is involved.

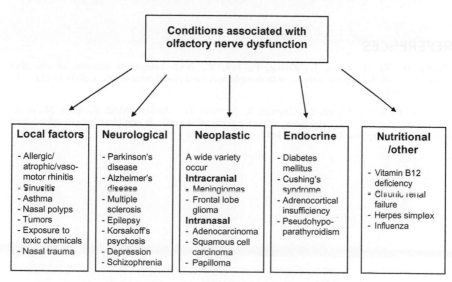

Local factors	Neurological	Neoplastic	Endocrine	Nutritional /other
- Allergic/ atrophic/vaso- motor rhinitis - Sinusitis - Asthma - Nasal polyps - Tumors - Exposure to toxic chemicals - Nasal trauma	- Parkinson's disease - Alzheimer's disease - Multiple sclerosis - Epilepsy - Korsakoff's psychosis - Depression - Schizophrenia	A wide variety occur **Intracranial** - Meningiomas - Frontal lobe glioma **Intranasal** - Adenocarcinoma - Squamous cell carcinoma - Papilloma	- Diabetes mellitus - Cushing's syndrome - Adrenocortical insufficiency - Pseudohypo- parathyroidism	- Vitamin B12 deficiency - Chronic renal failure - Herpes simplex - Influenza

Conditions associated with olfactory nerve dysfunction

Q:

Does the patient have to recognize the specific smell presented when testing to be classed as having a fully intact olfactory nerve system?

A:

No! The fact that the patient recognizes that an odor is being presented during testing is satisfactory. To be able to differentiate types of odors requires higher cortical function and is not relevant to determining if their olfactory nerves are working.

Q:

What is the current research related to the olfactory nerve?

A:

The olfactory nerve is very special. Derived from the embryonic nasal placode, the olfactory mucosa contains stem cells capable of regenerating. They give rise to olfactory neurons and olfactory ensheathing cells (OECs). By obtaining a nasal biopsy from a patient with a spinal cord injury, the OECs can be grown and purified. They then can be injected into the area of the spinal cord which is damaged. This pioneering research has resulted in patient's showing improvements in sensory and motor function (Huang et al., 2012; Tabakow et al., 2013).

REFERENCES

Huang, H., Xi, H., Chen, L., Zhang, F., Liu, Y., 2012. Long-term outcome of olfactory ensheathing cell therapy for patients with complete chronic spinal cord injury. Cell Transplant. 21 (Suppl. 1), S23–31.

Tabakow, P., Jarmundowicz, W., Czapiga, B., Fortuna, W., Miedzybrodzki, R., Czyz, M., et al., 2013. Transplantation of autologous olfactory ensheathing cells in complete human spinal cord injury. Cell Transplant. 22 (9), 1591–1612.

CHAPTER 2

Optic Nerve

THE ANATOMY—SUMMARY

The second cranial nerve is the optic nerve and measures about 4 cm in length. It is an unusual cranial nerve as it actually develops from the diencephalon. That means that the structures which receive and transmit the visual information are extensions of the forebrain and are therefore central nervous system (CNS) tracts formed by axons of the retinal ganglion cells. As such, and like the rest of the brain, the optic nerves are surrounded by the meninges (pia, arachnoid, and dura mater) and the cerebrospinal fluid (CSF) filled subarachnoid space.

THE ANATOMY—IN MORE DETAIL

Visual input reaches the posterior aspect of the eye and the pathway begins with the photoreceptive cells called cones and rods in the retina, adjacent to the pigment epithelium. The cone cells are what we need for color vision and function best in rather bright light. There are approximately 7 million cones in a human eye. The greatest density of cone cells is found within the *fovea centralis* of the *macula*. At this point, there is an absence of rod cells, and the further away from the fovea centralis, there is a sharp decline in numbers of cones and an increase in rod cells.

Cones have three different pigments within them depending if they are sensitive to long, short, or medium wavelengths of light. Long-wavelength cones, or L-cones, are sensitive to, and absorb, red light (564−580 nm). Short-wavelength cones, or S-cones, are sensitive to, and absorb, blue light (420−440 nm). Medium-wavelength cones, or M-cones, are sensitive to, and absorb, green light (534−545 nm). Therefore, in humans, we are known to have trichromatic vision. However, other vertebrates including birds, fish, and reptiles have four distinct cone photoreceptive cells. On development of night activity by placental mammals, two visual pigments were lost, with primates then developing the third visual pigment, resulting in the current state of trichromatic vision.

Clinical Anatomy of the Cranial Nerves. DOI: http://dx.doi.org/10.1016/B978-0-12-800898-0.00002-6

Rod cells on the other hand are significantly greater in number (130 million) and function in low light intensity. Most of these cells are found on the outer aspect of the retina and are important for peripheral vision. They are structurally similar to the cone cells and are able to be incredibly efficient at absorbing light and are therefore essential in night vision. Unlike cone cells, rod cells have only one light-sensitive visual pigment so do not really play a role in color vision.

The responses of the photoreceptor cells (first-order neurons) are transmitted to the bipolar cell layer (second-order neurons). These nerves have two processes—one coming from the rods and cones and the other projecting to the ganglion cell layer. From the ganglion cell layer, the impulses are then transmitted to the axons of the retinal ganglion cells. The axons of the retinal ganglion cells, which project centrally, are called the third-order neurons. This can be summarized in the diagram below.

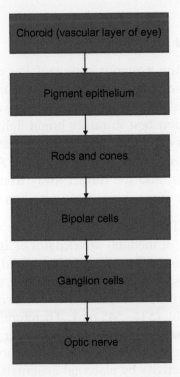

The central processes of the axons of the retinal ganglion cells join to form the optic nerve. The optic nerve passes backward and in a medial direction. It passes through the posterior part of the orbital cavity, running through the optic canal (Figure 2.1).

Within the orbit, the optic nerve is closely related to the four recti muscles. The ciliary ganglion (from the oculomotor nerve) is found between the optic nerve and the lateral rectus.

As the optic nerve passes through the small *optic canal* (0.5 cm long), it contains the layers of the meninges, including the CSF-filled subarachnoid space. The central artery (from the ophthalmic artery) and vein of the retina travel through the meninges, passing to the anterior part of the optic nerve.

Within the cranial cavity, each optic nerve passes posterior and medially for approximately 1 cm to the *optic chiasma*. From this point, nerve fibers on the medial, or nasal side of each retina, cross the

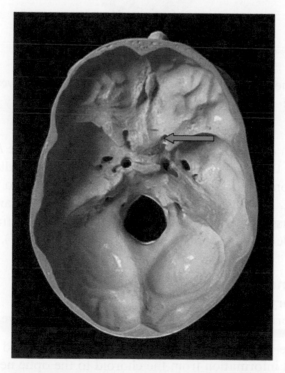

Figure 2.1 The green arrow highlights the position of the optic canal on the base of skull. This is the point of exit of the optic nerve from the orbit.

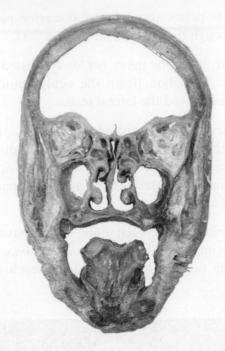

Figure 2.2 An unlabeled image to aid identification of the optic nerve.

midline to the optic tract of the opposite side. However, the nerve
fibers on the lateral, or temporal side, pass posteriorly but do not
cross.

From each *optic tract*, visual information passes to the *lateral geniculate body* in the *midbrain*. From this point, the optic radiation carries
visual input onward to the optic radiation and then on to the occipital
visual cortex. A small number of fibers for ocular and pupillary
reflexes bypass the lateral geniculate body and pass straight to the
pretectal nucleus and superior colliculus (Figures 2.2–2.7).

The Components
Special Sensory
• Visual information. It is a special visceral afferent nerve.

Within the retina, there are 10 layers of cells involved in transmission of visual information from the choroid to the optic nerve:

1. Pigment cell layer
2. Layer of cones and rods

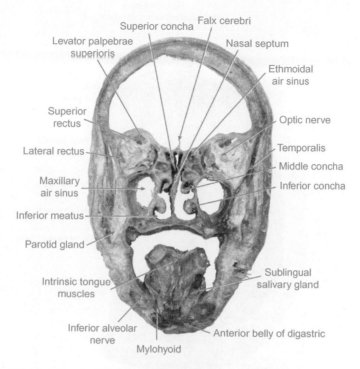

Figure 2.3 A fully labeled image of the position of the optic nerve in relation to other structures.

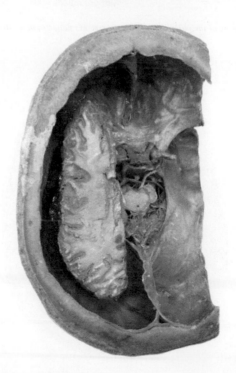

Figure 2.4 An unlabeled image to aid identification of the optic nerve.

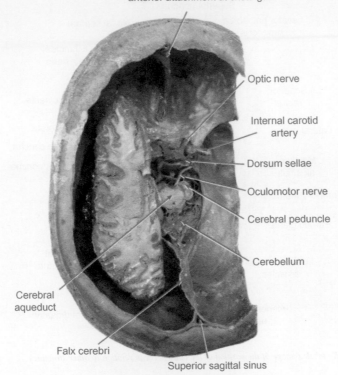

Falx cerebri—
anterior attachment at crista galli

Optic nerve

Internal carotid
artery

Dorsum sellae

Oculomotor nerve

Cerebral peduncle

Cerebellum

Cerebral
aqueduct

Falx cerebri

Superior sagittal sinus

Figure 2.5 A fully labeled image of the position of the optic nerve in relation to other structures.

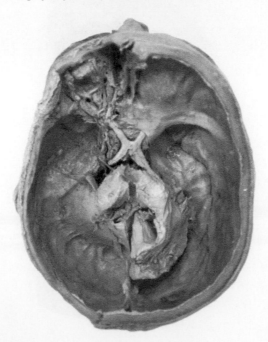

Figure 2.6 An unlabeled image to aid identification of the optic nerve.

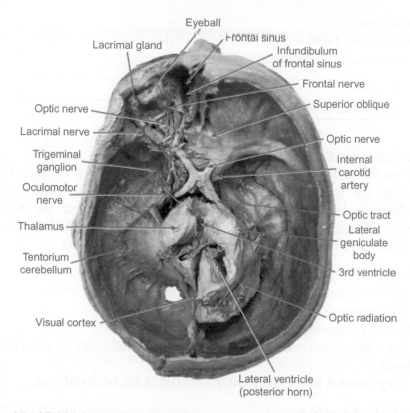

Eyeball
Lacrimal gland
Frontal sinus
Infundibulum of frontal sinus
Frontal nerve
Superior oblique
Optic nerve
Lacrimal nerve
Trigeminal ganglion
Oculomotor nerve
Thalamus
Tentorium cerebellum
Visual cortex
Optic nerve
Internal carotid artery
Optic tract
Lateral geniculate body
3rd ventricle
Optic radiation
Lateral ventricle (posterior horn)

Figure 2.7 A fully labeled image of the position of the optic nerve in relation to other structures.

3. Outer limiting membrane
4. Outer nuclear layer
5. Outer plexiform layer
6. Inner nuclear layer
7. Inner plexiform layer
8. Ganglion cell layer
9. Nerve fiber layer
10. Inner limiting membrane

From the ganglion cells, these form the optic nerve. The pathway of visual information from the optic nerve to the brain is summarized below.

Optic nerve → Optic chiasma → Optic tract → Lateral geniculate body → Occipital visual cortex

The Ganglia

The following table summarizes the ganglion associated with the optic nerve, including its location and function.

Ganglion	Location	Function
Retinal ganglion cells	Retina	Special sensory ganglion and cell bodies for vision

The Important Branches

The retina and its projections can be divided up by a circle halved through the fovea based on its projections from either the nasal (medial) or temporal (lateral) halves. The lens inverts the visual images. Therefore, in the left eye, the temporal half sees the left side of the world and the nasal half sees the right side of the world.

Left Eye—Temporal Side
1. Visual information passes to the right side of the left retina.
2. This passes to the left optic nerve.
3. It then crosses the optic chiasm.
4. The visual information then passes to the right lateral geniculate body.
5. Then onward to the right visual cortices of the occipital lobe.

Left Eye—Nasal Side
1. Visual information passes to the left side of the left retina.
2. This passes to the left optic nerve.
3. NO crossing over of this information occurs at the optic chiasm.
4. The visual information then passes to the left lateral geniculate body.
5. Then onward to the left visual cortices of the occipital lobe.

Right Eye—Temporal Side
1. Visual information passes to the left side of the right retina.
2. This passes to the right optic nerve.
3. It then crosses the optic chiasm.
4. The visual information then passes to the left lateral geniculate body.
5. Then onward to the left visual cortices of the occipital lobe.

Right Eye—Nasal Side
1. Visual information passes to the right side of the right retina.
2. This passes to the right optic nerve.

3. NO crossing over of this information occurs at the optic chiasm.
4. The visual information then passes to the right lateral geniculate body.
5. Then onward to the right visual cortices of the occipital lobe.

THE CLINICAL APPLICATION

Tip!

When examining a patient's eye for visual acuity and fields, ask if they normally wear glasses and if they are short- or long-sighted. It would be embarrassing to carry out these tests without the patient being able to correct a visual defect themselves, which was already established from the patient's history. In addition, the same holds true if they normally wear contact lenses and do not have them in when they normally would! It must also be tested without the aid of glasses or contact lenses to identify if their visual defect has worsened or changed in any way.

There are five areas that you want to test when testing the optic nerve and its functions. These are as follows:

1. Visual acuity
2. Visual fields
3. Pupil size (reflexes)
4. Color assessment
5. Ophthalmoscopy

Visual Acuity

This is a measure of central vision and tests the ability of the patient to identify shapes and objects. Each eye MUST be tested separately. If a patient normally wears glasses or has contact lenses, perform the examination with these, and then without to ensure the patient's visual defect has been examined to identify any changes or worsening of their original condition.

A pocket visual acuity chart can be used for a general assessment, or a Snellen's Test Type chart can be used for more formal testing. This chart comprises different sized letters and is used to test DISTANT VISION. The Snellen chart is placed at 20 ft in the United States or 6 m away from the patient in all other countries. The point where a person with normal visual acuity can read to at 20 ft/6 m is referred to as 20/20 vision (United States) or 6/6 vision (rest of the world). Therefore, a score for each eye has to be recorded when undertaking assessment of visual acuity.

Having 20/20 or 6/6 vision does not automatically mean perfect vision as it only tests central vision and does not take into account peripheral visual fields, color awareness, or depth of perception.

In addition to this, near vision must also be examined. This is undertaken by using a card with sentences of different font sizes. Each paragraph has "points" where they are 1/72 in. apart. Therefore, N48 is the largest of the types and N5 is the smallest. This chart should be read at a comfortable reading distance of approximately 35 cm from each eye.

Visual Fields

This part of the examination of the optic nerve allows the examiner to assess both central and peripheral vision.

Confrontation Visual Field Testing (Donder's Test)

This involves standing or sitting (both patient and examiner) at the same eye level.

1. Ask the patient to cover one eye, for example, their right eye (examining the patient's left eye).
2. Ask the patient to remain looking at your eyes and to say "now" when the examiner's finger enters from out of sight into their peripheral vision.
3. Once the patient understands this, the examiner should cover their left eye with their left hand (opposite side to the patient).
4. Beginning with the examiner's hand and arm fully extended, slowly bring your outstretched fingers centrally.
5. The examiner should bring their hand and outstretched fingers centrally.
6. The patient should then say "now" at the same time they see the examiner's outstretched fingers and hand entering their field of vision.

7. This procedure should be undertaken once every 45° out of the 360° of peripheral vision.

Tip!

The examiner should wiggle their index and middle finger to allow the patient to detect movement of those digits when bringing their hand and outstretched hand in from the periphery at each of the 45° out of the 360° circle.

Tip!

When examining the nasal side of the visual fields, change hand covering the examiner's eye, that is, if examining the patient's nasal side of their left eye, the examiner should switch to their right hand covering their left eye. That way, it ensures that the examiner's hand does not cross the visual fields yielding a false-positive result. On examining the temporal field of the patient's left eye, the examiner should cover their left eye with their left hand therefore using their right hand to conduct the examination on that half of the 180° circle.

Perimetry Test

This is a formal way of recording the visual fields. It systematically tests light sensitivity and identification by the patient on a predefined background.

Nowadays, this test is automated. The patient would sit down and look into a bowl-shaped instrument (a perimeter). The patient then has to stare at the center of the bowl and identify flashes of light. Each time the light flashes at varying points in the visual field (both central and peripheral), the patient presses a button to record when they see the flash of light.

A computer records when and where the light flashed and when the patient pressed the button indicating when they saw the light. A printed image is then obtained as to where the patient did not record seeing the visual flashing light. This then allows mapping of where the potential visual field defect lies.

Older forms of these tests include the Goldmann perimeter or tangent screen tests where the examiner would have to be involved in manual recording or stimulation. Also, perimetry assessment used to be tested with either static perimetry testing or kinetic perimetry testing (like the Goldmann test). These have now largely been replaced with the automated system.

Pupil Size (Reflexes)

The pupils are approximately equal in size varying from 2–4 mm (bright light) to 4–8 mm (dark light). When a light is shone into one eye, the pupil will constrict and this is referred to as the *direct response*. The pupil opposite to the side the light is shone into also constricts and this is referred to as the *indirect response*.

One acronym that is popular in clinical notes when documenting the pupil is PERRLA. This abbreviation stands for:

Pupils
Equal
Round and
Reactive to
Light and
Accommodation

Although this provides a simple overview of the pupil, it fails to take into account the actual size and shape of the pupil, symmetry (or lack of), and the response to the pupil constricting.

Tip!

As much as possible, examine the pupils in a dimly lit room. If you are in a ward or an accident and emergency room environment, it is usually possible to switch off the light within the examining area.

Ask the patient to fix their gaze on something in the distance as if they try to focus on you or the examining light, reflex miosis (constriction of the pupil) will occur making the examination difficult to conduct.

Testing at the Bedside

- Ask the patient to focus on something in the distance, perhaps on the other side of the room.
- Record the approximate diameter and shape of the pupil.
- SLOWLY move the light to the patient's eye from below the level of the eye, checking the pupillary response on the side the light is shone into, as well as the opposite one.
- Grade each pupil's reaction to light from +1 to +4, with +4 being the fastest reaction.
- Then, ask the patient to focus from far to a near object, that is, if they were focussing on something at the opposite side of the room, ask them then to focus on, for example, a pen tip approximately 20 cm away from them.
- Again, grade the pupillary response to this. The normal response is for miosis (pupillary constriction) when changing focus from a far to near object. It should be recorded in the same way as for the reaction to light (i.e., +1 to +4).
- An example as to how to record the pupil in your clinical examination is shown below.

Pupillary Assessment

Pupil	Left	Right
Shape	Round	Round
Size	6 mm	7 mm
Light	+4	+4
Near object focus	+4	+4

Two iris muscles control the diameter of the pupil: (i) *sphincter muscle* (also called the pupillary constrictor or sphincter pupillae) and (ii) *dilator muscle* (radial muscle of iris). The sphincter muscle is controlled by the parasympathetic nervous system and the dilator muscle is controlled by the adrenergic sympathetic nervous system and also circulating catecholamines.

For pupillary constriction, as mentioned earlier, the pathway of this involves not just the optic nerve but also the third cranial nerve, which is called the oculomotor nerve. The following diagram summarizes the pathway when light is shone into the pupil.

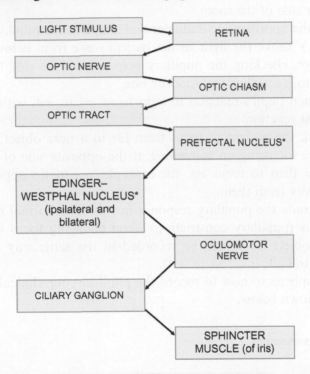

On the diagram, where * is shown at the pretectal and the Edinger-Westphal nuclei, this is the point where axons from each pretectal nucleus pass ipsilaterally and contralaterally to the ipsilateral and contralateral Edinger-Westphal nuclei. It is this point in the pathway that allows for constriction of the pupil on the side light is shone into the eye, but also the opposite eye (despite no light being shone into it).

The ciliary ganglion, therefore, is where the parasympathetic fibers will synapse, before the impulses are conveyed to the sphincter muscle of the eye via the short ciliary nerves.

Color Assessment

This is not typically assessed on the ward as a routine procedure. Color vision deficiency is typically referred to as color blindness. Complete color vision deficiency where the individual is unable to determine any colors is rare. Inherited color vision deficiency affects men more than women with approximately 8% of men and 0.5% of women affected by it. Interestingly, there are a variety of causes that can result in acquired color vision deficiency (e.g., chemical exposure, head injuries, or with a wide range of medications). This occurs in approximately 5% of the population (Guidance Note MS7, HSE). Indeed, subtle differences in color vision perception occur as we age. Typically, there are two versions of color vision deficiency:

1. Red-green deficiency: this is the most common of the inherited types and the patient is unable to differentiate shades of red and green.
2. Blue-yellow deficiency: this rare condition affects the ability to differentiate blue and green, with yellow appearing as a pale purple or grey.

An optometrist can carry out this test. There are many types of tests which can be used to assess color vision. The most common type to be used is the *Ishihara test*. It consists of multiple plates (24 or 38) made up of multicolored dots. The patient is asked to look at each plate in turn and state the number that they recognize.

Recently, the City University Test is a two-part test (The City University Colour Vision Test, UK). The initial part is a screening aid with the second part assessing the severity of the color defect. Charts are used for the first part with part two using displays of a central color with four peripheral colors. The patient being examined then has to select one of the outer colors closest to the center color.

For those careers that require absolute color recognition (e.g., those in the aviation industry like pilots, maritime, naval, railways, and air forces), lantern testing can also be used where it imitates the actual signal systems that those posts would encounter.

Ophthalmoscopy

The retina is the only part of the CNS which can be visualized from the external environment. As the eye is an extension of the brain, and as such carries its external layers of the meninges, any pressure increases within the cranial cavity can be transmitted straight along the optic nerve and visualized with an ophthalmoscope.

Tip!

As well as a darkened room, it may help to instil mydriatic drops to dilate the pupil. Please check your local protocols and procedures but this may include one to two drops of 0.5% tropicamide. This should be used approximately 15–20 min prior to examining the eye.

Tip!

Use the right hand and eye when examining the patient's right eye and vice versa for the left. In examining the retina you should "pivot" with the ophthalmoscope tilting it up and down and left to right.

The examination should consist of the following:

1. Explain to the patient everything that you want to do. This will aid compliance.
2. Examine the lens and vitreous approximately 1 m away. Use high + numbers on the ophthalmoscope.
3. Check for the red reflex which is the red glow from the choroid.
4. Cataracts will be seen as a black pattern obstructing this reflex.
5. Blood or loose floaters in the vitreous will be identified as black floaters.
6. Identify the position of the opacity in the eye.
7. When the retina is in focus, the optic disc should be examined.

Tip!

When examining the opacity in the lens, the examiner should move their head in the horizontal plane when using the ophthalmoscope. If the opacity moves in the same direction as the examiner's gaze, the opacity lies behind the lens. If the opacity moves in the opposite direction to the examiner's gaze, the opacity lies in front of the lens. If the opacity does not move at all, the pathology is within the lens.

Tip!

Normal optic cups are similar on the left and right sides and occupy <50% of the optic disc. An increase in intraocular pressure (i.e., >21 mmHg) results in cupping of the optic disc, capillary closure, and ultimately nerve damage with sausage-shaped field defects, which are called scotomata. These occur near the blind spot which if these unite, can result in a major visual field defect.

PATHOLOGIES

Numerous pathologies can affect the eye and the optic nerve. Here, only a selection of some of the most important and/or serious conditions that affect this nerve are presented.

Tip!

If a patient has a sudden loss of vision, specialist advice should be sought immediately.

Refractive Errors
Myopia—Short Sighted
Myopia is where the eyeball itself is too long in an antero-posterior direction. This means that the light is not focussed on the retina but rather falls in front of the retina. In the normal eye, when an object comes closer to the eye, the further back the image falls. With a myopic eye, the closer object will therefore focus on the retina. To help focus the eye when viewing far away objects, a concave glass or contact lens is required.

Hypermetropia—Long Sighted
Hypermetropia is where the eye is too short in the antero-posterior direction. When the eye is at rest, the objects further away will be focussed behind the retina. When this happens, the ciliary muscles contract to allow the lens to become more convex, helping it to focus on the object. However, this will result in the eye becoming tired and can result in a convergent squint in children. With hypermetropia, the correction is undertaken with convex spectacle lenses. This helps bring the image forward to allow it to focus on the retina.

Visual Field Defects
The most important things which have to be assessed when assessing for visual field defects are if it is a unilateral or bilateral problem, if

clear boundaries exist, and if so where they exist in the vertical or horizontal planes, and where.

A simple rule to follow is in relation to the optic nerve pathway as follows:

Site of Lesion	Resulting Visual Field Defect
Anterior to the optic chiasm	Loss of vision on same side as damage
Optic chiasm	Bitemporal hemianopia (though can be complicated dependent on the fibers involved)
Posterior to the optic chiasm	Loss of vision in visual field opposite to damage

The difficulty is trying to locate the exact location of the visual field defect. One thing to keep in mind is that the upper parts of the visual field pass to the lower part of the retina (and also vice versa). Fibers from the nasal half of the retina cross over in the optic chiasm to then join with the uncrossed temporal retinal fibers. Therefore, a pituitary tumor which is pressing on the optic chiasm will press on the nasal fields from the retina resulting in a bitemporal hemianopia. However, if that tumor grows more to one side, there will also be a superimposed central optic nerve defect as well as the bitemporal hemianopia which will complicate the presentation when formally examined. A contralateral superior homonymous quadrantanopia could be caused by a temporal lobe tumor. Posterior visual cortex lesions result in nonperipheral homonymous hemianopia. If there is a homonymous hemianopia, this can be from something affecting the entire optic radiation or visual cortex.

The Red Eye
The red eye is a common condition to present to accident and emergency and general practitioners. They are painful but can also be dangerous to vision and require urgent specialist input from the ophthalmologist (e.g., due to acute glaucoma, corneal ulceration, or acute iritis).

If the red eye is due to acute glaucoma, it generally presents as blurred vision or haloes around lights especially during the night. It is caused by blockage at the Canal of Schlemm resulting in an inability of aqueous to drain from the anterior chamber. It is treated by a drug to cause miosis (e.g., pilocarpine), which will open the blockage, by improving the drainage angle. Also a drug to reduce the formation of aqueous can also be given (e.g., acetazolamide).

The uvea is the vascular and pigmented part of the eye and includes the choroid, iris, and ciliary body. Collectively, the iris and ciliary body are referred to as the anterior uvea, with iritis typically involving the ciliary body too, hence the phrase anterior uveitis. The main aim of treatment is prevention of damage to the eye and steroids and a drug to help with any adhesions could be used (e.g., cyclopentolate). These types of problems should be discussed with a specialist, especially if there is any doubt with the clinician.

Optic Neuritis

Optic neuritis is inflammation of the optic nerve and can result in complete or partial loss of vision. Most commonly, optic neuritis happens to patients who have multiple sclerosis (MS). Indeed, it may be the presentation of optic neuritis that is the first indication that someone has MS. A variety of presentations in disturbed vision can occur from blurring through to blindness. Most cases resolve spontaneously.

Papilloedema

Optic disc swelling can be due to a number of conditions, with papilloedema referring to swelling of the optic disc due to raised intracranial pressure (ICP). If papilloedema is present, there is generally no difficulty in identifying it. It presents with an engorged optic disc without clear margins and hemorrhaging can be present.

Papilloedema is a medical emergency, and NO patient should be discharged without exclusion of a space occupying lesion or venous sinus thrombosis. It is imperative that the patient's visual fields are also examined and discussion is undertaken urgently with a neurologist and/or neurosurgeon for immediate neuroimaging.

INTERESTING CLINICAL QUESTIONS

Q:

What is the typical pattern on visual field loss caused by a pituitary tumor?

A:

Generally, pituitary tumors are not common and tend to be benign (pituitary adenoma). The type of visual field loss obviously depends on where the pituitary adenoma extends. Typically, with suprasellar extension, the typical visual field loss is a bitemporal hemianopsia. Information from the nasal retinae are affected and as such, as it carries information from the temporal fields, pressure by a pituitary adenoma results in loss of the peripheral visual fields. The patient will say they have tunnel vision. They will complain of difficulty with their peripheral fields especially when driving.

Q:

Why can an increase in ICP (for whatever reason) be observed in using the ophthalmoscope?

A:

The optic nerve is classically described as an outgrowth of the brain. As such it carries with it its own meningeal layers, as surround the brain and spinal cord (i.e., the dura, arachnoid, and pia mater). Any increase in ICP will result in transmission of this raised pressure straight along the optic nerve and its surrounding meninges. This results in the clinical finding of papilloedema.

REFERENCES

Guidance Note MS7, third ed., Health and Safety Executive. <http://www.hse.gov.uk/pubns/ms7.pdf> (accessed 06.12.13.).

The City University Colour Vision Test, third ed., Keeler 1998.

CHAPTER 3

Oculomotor Nerve

THE ANATOMY—SUMMARY

The third cranial nerve is the oculomotor nerve. After it emerges from the midbrain, it passes to the orbit by passing through the superior orbital fissure. It not only supplies the majority of the extraocular muscles but also carries parasympathetic fibers in it for pupillary constriction. In addition, it carries fibers that also maintain the eye open via the levator palpebrae superioris muscle.

THE ANATOMY—IN MORE DETAIL

The oculomotor nerve arises from the anterior surface of the midbrain. Its nucleus is ventral to the aqueduct. From here it passes through the dura mater to pass between the superior cerebellar and posterior cerebral arteries. From here, it passes to the middle cranial fossa in the lateral wall in the cavernous sinus. At this point, it also receives a few filaments from the cavernous plexus of the sympathetic nervous system, as well as a small communicating branch from the trigeminal nerve. Just as it enters the superior orbital fissure (see Figure 3.1) to gain access to the orbit, it divides into a superior and inferior division. The superior division innervates the levator palpebrae superioris and the superior rectus. The inferior division innervates the inferior oblique and the medial and inferior recti. The oculomotor nerve also carries fibers in it for parasympathetic innervation (or visceral efferent) via the ciliary ganglion to the *sphincter* (also called the pupillary constrictor or sphincter pupillae) and ciliary muscles (Figures 3.2 and 3.3).

The control of eye movements can be subdivided as follows:

1. Central upper motor neurons
2. Oculomotor, abducens, and trochlear nerves and muscles they supply.

Clinical Anatomy of the Cranial Nerves. DOI: http://dx.doi.org/10.1016/B978-0-12-800898-0.00003-8

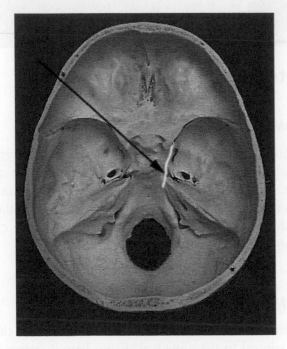

Figure 3.1 The oculomotor nerve is highlighted in yellow as it passes anteriorly to enter the orbit at the superior orbital fissure. The arrow indicates the position of the oculomotor nerve as it passes close to the posterior clinoid process.

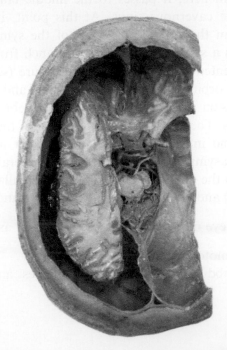

Figure 3.2 An unlabeled image to demonstrate the position of the oculomotor nerve in relation to other structures.

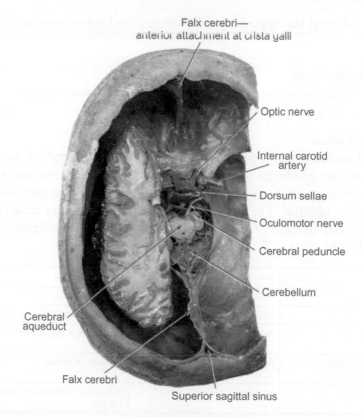

Falx cerebri—
anterior attachment at crista galli

Optic nerve

Internal carotid
artery

Dorsum sellae

Oculomotor nerve

Cerebral peduncle

Cerebellum

Cerebral
aqueduct

Falx cerebri

Superior sagittal sinus

Figure 3.3 A labeled image to demonstrate the position of the oculomotor nerve in relation to other structures.

Conjugate Gaze

The fast voluntary and reflex eye movements have their origins in the frontal lobe. These fibers then pass via the internal capsule, crossing in the pons and ending into the paramedian pontine reticular formation (PPRF) which is the center for lateral gaze. It also receives fibers from

1. ipsilateral occipital cortex—responsible for tracking or pursuing objects in the visual fields
2. vestibular nuclei—for eye movements and position of the head.

Lateral eye movements are controlled through the center of lateral gaze via the medial longitudinal fasciculus. Fibers from here pass to the ipsilateral abducens nerve and after crossing the midline, the opposite oculomotor nerve nucleus. Therefore, each sixth nerve nucleus and the opposite third nerve nucleus are linked, i.e., the lateral rectus on one side and the opposite medial rectus, and other muscles detailed later.

The Components

The oculomotor nerve is comprised of two different components: a somatic motor and visceral motor division.

Somatic Motor

The cell bodies of the somatic motor part of the oculomotor nerve are found in the midbrain and they supply motor innervation to the medial, inferior, and superior recti, levator palpebrae superioris, and the inferior oblique. The following table summarizes the attachments and functions of each of these muscles in relation to movement of the eyeball/eyelid.

Muscle	Attachments	Actions
Medial rectus	Annular tendon → eyeball	Adduction of eyeball
Inferior rectus	Annular tendon → eyeball	Depression and adduction of eyeball
Superior rectus	Annular tendon → eyeball	Elevation and medial rotation of eyeball
Levator palpebrae superioris	Sphenoid bone → upper tarsal plate	Elevation of eyeball
Inferior oblique	Maxilla → eyeball	Abduction, elevation, and lateral rotation

Visceral Motor

The visceral motor component of the oculomotor nerve supplies parasympathetic innervation (or visceral efferent) via the ciliary ganglion to the *sphincter* (also called the pupillary constrictor or sphincter pupillae) and ciliary muscles. Therefore, the presynaptic part is found in the midbrain with the postsynaptic fibers originating in the ciliary ganglion. The effect of stimulation of the visceral motor part of the oculomotor nerve results in constriction of the pupil and accommodation of the lens.

The Nuclei

Two nuclei are associated with the oculomotor nerve—the oculomotor nucleus and the Edinger–Westphal nucleus.

1. Oculomotor nucleus—This supplies all the above-named muscles in the table. Therefore, it supplies ALL the extraocular muscles apart from the superior oblique and the lateral rectus.

2. Edinger–Westphal nucleus—This nucleus contains the parasympathetic fibers for pupillary constriction sphincter pupillae and accommodation (ciliary muscles).

The Important Branches

There is an easy way to remember what the oculomotor nerve actually supplies in relation to the extraocular muscles by the following tip box.

Tip!

The following mnemonic will aid to remembering the supply of the extraocular muscles. It looks like a sulfur compound that we used to encounter in school chemistry.

$$LR_6SO_4$$

It simply means:

Lateral rectus—supplied by the sixth cranial nerve (Abducens nerve)
Superior oblique—supplied by the fourth cranial nerve (Trochlear nerve).

ALL OTHER extraocular muscles are supplied by the oculomotor nerve (as described in the table earlier).

THE CLINICAL APPLICATION

Testing at the Bedside

There are two aspects to the clinical testing of the oculomotor nerve: eye muscle movement and the pupillary reflex.

Tip!

ALWAYS explain to the patient what you are going to do and what you want them to do. That way compliance will be encouraged and a better level of trust will be built up between patient and clinician.

Extraocular Eye Muscle Testing

This is used to test the oculomotor, trochlear, and abducens nerves (as all of these nerves supply the extraocular muscles and, therefore, are involved in movement of the eyeball).

1. Ask the patient to keep their head still during the examination
2. With their eyes only, they should follow the tip of your finger (or pen torch, or pencil, etc.)
3. The examiner should then move the object in the horizontal plane from extreme left to extreme right
4. This should be done SLOWLY
5. When at the extreme left or right, with the examining object, stop!
6. Observe for nystagmus
7. Then an H-shaped pattern with the examining object can be drawn in space, which the patient should follow WITH THEIR EYES ONLY
8. All movements should be done SLOWLY to assess for eye movements.

Pupillary Reflex
This can be tested in two ways:

1. Ask the patient to follow your finger as you move it closer to their eyes. This will induce accommodation and therefore, pupillary constriction.
2. The swinging flashlight test can be used. This involves shining a pen torch into each pupil and observing both pupils for pupillary constriction. The purpose of this test is to identify if a relative afferent pupillary defect (RAPD) is present (discussed later).

Advanced Testing
Abnormality in these initial tests or continuing clinical suspicion of oculomotor nerve damage should result in further investigation to identify any pathology which may be present. It may be necessary at this point to consult specialists for advice.

Further testing can be undertaken by examining brainstem nuclei which provide input to the oculomotor nerve. This would involve examining the patient's five aspects of ocular function namely fixation, saccadic movements, pursuit movements, compensatory movements, and opticokinetic nystagmus (Walker et al., 1990).

In terms of fixation, saccadic movements and pursuit movements, observing the eye position and its related movements can assess these elements. For the compensatory movements, nystagmus should be assessed, as well as the doll's head maneuver.

Doll's Head Maneuver (Caloric Testing/Vestibular Caloric Stimulation)
The doll's head maneuver is also referred to as caloric testing or vestibular caloric stimulation. It is a test of the vestibule-ocular reflex and can be used to diagnose pathology in the peripheral vestibular system via eye movements. It is also a test used in brainstem death testing. Therefore, it tests numerous structures namely the labyrinthine structures, pontine lateral gaze centers, medial longitudinal fasciculus (crossed fiber tract which carries information about which direction the eyes should move connecting the oculomotor, trochlear, and abducens nerves), and nuclei and peripheral nerves of the oculomotor and abudcens nerves. It should be tested as follows:

1. ALWAYS inform the patient what you are going to do. Take time to explain this test, as it can be slightly unpleasant.
2. Warm water ($>44°C$) or cold water can be used. It may be easier in the clinical environment obtaining colder water.
3. ALWAYS check the integrity of the tympanic membrane first to ensure there is no perforation present.
4. Elevate the patient's head to approximately 30°. This ensures that only the horizontal semicircular canals will be stimulated.
5. Inject the cold (or warm water) into the auditory tube. In the conscious patient <1 ml may stimulate the response, whilst it may take up to 50 ml in the unconscious patient. This should only be done for <1 min.
6. The eyes should be observed over up to 3 min.

Response—This depends if the patient is conscious or not.

Conscious Patient

It depends if hot or cold water is used as the initial stimulus

There are two components of nystagmus—fast and slow.

Cold Water
The slow component of nystagmus is *toward* the side of injected cold water.

The fast component of nystagmus is then *away* from the side of injected cold water.

Warm Water

The slow component of nystagmus is *away* from the side of injected hot water.

The fast component of nystagmus is *toward* the side of injected hot water.

Tip!

There is a mnemonic to aid remembering which direction the eyes move in the *fast* component of nystagmus when performing the caloric test.

COWS

C—Cold
O—Opposite
W—Warm
S—Same

Unconscious Patient

Provided that the patient has a normal vestibular system and no brainstem pathology, no fast nystagmus is present, that is, if cold water is injected, there is only movement of the eye to the ear which has been injected with cold water.

PATHOLOGIES

Normally the oculomotor nerve allows for the following eye movements:

1. Elevation when adducted (inferior oblique)
2. Elevation when abducted (superior rectus)
3. Depression when abducted (inferior rectus)
4. Adduction (medial rectus)
5. Eyelid elevation (levator palpebrae superioris)
6. Pupillary constriction (sphincter muscle).

If there is a complete paralysis of the oculomotor nerve, with intact trochlear and abducens nerves, the following will be seen:

1. Eye "down and out," i.e., due to the intact functions of the trochlear (lateral rotation and depression of eye) and abducens nerves (abducts the eye)
2. Pupillary dilation and unresponsive (paralysis of sphincter muscle)
3. Ptosis (droopy eyelid due to paralysis of levator palpebrae superioris).

For diagnosing oculomotor nerve lesions, Warwick's scheme for the nerve complex can be used (Warwick, 1953). From the clinical signs, a potential point of nerve damage of the oculomotor nerve could be identified before more in-depth investigation, e.g., by CT/MRI scanning.

1. Conditions not representing an oculomotor nerve nuclear complex pathology (all unilateral presentations)
   a. External ophthalmoplegia (weakness or paralysis of an extraocular muscle) ± pupillary involvement (with a normal contralateral superior rectus function)
   b. Internal ophthalmoplegia
   c. Ptosis.
2. Conditions that can potentially be oculomotor nerve nuclear complex in origin
   a. Complete bilateral oculomotor nerve paralysis
   b. Ptosis (bilateral)
   c. Internal ophthalmoplegia (bilateral)
   d. Medial rectus paralysis (bilateral)
   e. Singular muscle involvement (NOT superior rectus or levator palpebrae superioris).
3. Oculomotor nerve nuclear complex lesion
   a. Ipsilateral oculomotor nerve palsy and contralateral superior rectus paralysis with partial ptosis bilaterally
   b. Bilateral oculomotor nerve paralysis (±internal ophthalmoplegia).

Also, dependent on the pathology, the following classification can also be used clinically.

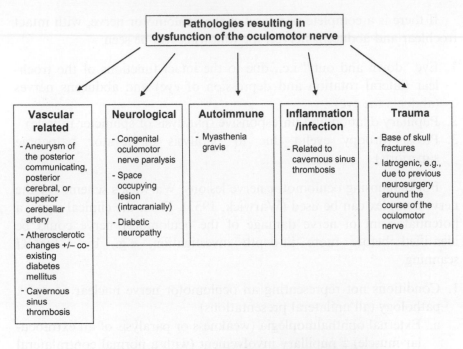

		Pathologies resulting in dysfunction of the oculomotor nerve		
Vascular related	**Neurological**	**Autoimmune**	**Inflammation /infection**	**Trauma**
- Aneurysm of the posterior communicating, posterior cerebral, or superior cerebellar artery - Atherosclerotic changes +/– co-existing diabetes mellitus - Cavernous sinus thrombosis	- Congenital oculomotor nerve paralysis - Space occupying lesion (intracranially) - Diabetic neuropathy	- Myasthenia gravis	- Related to cavernous sinus thrombosis	- Base of skull fractures - Iatrogenic, e.g., due to previous neurosurgery around the course of the oculomotor nerve

INTERESTING CLINICAL QUESTIONS

Q:

What signs would you notice in a true oculomotor nerve paralysis, and why would you see those signs?

A:

Complete paralysis of the oculomotor nerve will affect the majority of the ocular musculature. There will be a ptosis (drooping upper eyelid) due to loss of supply to the levator palpebrae superioris. The unopposed action of orbicularis oculi (supplied by the facial nerve) will result in a contributing factor to the ptosis. The eye will have the classic pattern of appearing "down and out." This is due to the unopposed actions of the superior oblique (which is supplied by the trochlear nerve) and the lateral rectus (which is supplied by the abducens nerve). In addition, the pupil is fully dilated as there is unopposed action of the sympathetic nervous system on the dilator muscle of the eye.

Q:
With a pathology affecting the oculomotor nerve, when might the pupil be spared, i.e., pupillary dilation NOT typically be seen despite other oculomotor nerve pathology related signs (e.g., ophthalmoplegia)?
A:
It may be the case that the patient has small vessel disease, e.g., diabetes mellitus or hypertension, resulting in a palsy of the oculomotor nerve. When this occurs, pain can be significant, but generally resolves in a few days. Pathologies are typically found in the subarachnoid region or within the cavernous part of the pathway of the oculomotor nerve.

However, not all patients have a vascular-related condition as lesions found at the cavernous sinus or at the point of entry of the oculomotor nerve into the orbit (superior orbital fissure) may cause oculomotor nerve pathology with pupil sparing signs. It may be that the inferior part of the oculomotor nerve is spared and the fibers destined for the sphincter muscle, in the process of joining the twig to the inferior oblique (also carried in the inferior division) may take a separate course from that point. The fibers then for the sphincter muscle carried in the inferior division (closely related to the inferior oblique fibers) may pass more lateral or inferior, or pass deeper in the substance of the nerve, thus assuming a more protected position. Therefore, this would preserve the sphincter pupillary muscle, but still affect the vast majority of other muscles that the oculomotor nerve supplies, as previously described.

Q:
What is a Holmes-Aide pupil?
A:
A Holmes-Aide pupil is a neurological condition affecting the pupillary size and response to light. Typically, it results in a larger pupil

which reacts slowly (or not at all) to light with absent tendon jerk reflexes, especially of the Achilles tendon. It tends to be because of a viral infection affecting the ciliary ganglion and the dorsal root ganglion in the spinal cord. It is not life threatening and symptomatic treatment may be the best option.

REFERENCES

Walker, H.K., Hall, W.D., Hurst, J.W., et al., 1990. Clinical Methods, The History, Physical, and Laboratory Examinations, third ed. Butterworths, Boston, ISBN-10: 0-409-90077-X.

Warwick, R., 1953. Representation of the extraocular muscles in the oculomotor nuclei of the monkey. J. Comp. Neurol. 98, 449–503.

CHAPTER 4

Trochlear Nerve

THE ANATOMY—SUMMARY

The fourth cranial nerve is the trochlear nerve. The trochlear nerve is a rather unusual cranial nerve in comparison to the others because:

- It is the smallest of the cranial nerves
- Longest intracranial course of the cranial nerves
- Only one of the cranial nerves to arise from the dorsal aspect of the midbrain
- One of only two nerves to decussate (the other is the optic nerve).

It supplies a single muscle—the superior oblique.

THE ANATOMY—IN MORE DETAIL

As a long slender structure, the trochlear nerve arises from the posterior (dorsal) aspect of the midbrain. It lies immediately caudal to the nucleus of the oculomotor nerve. The fibers of the trochlear nerve pass around the periaqueductal grey matter decussating at the level of the superior medullary velum. This is a small area of white matter passing between the two superior cerebellar peduncles. It forms part of the roof of the fourth ventricle.

The trochlear nerve exits the midbrain just below the level of the inferior colliculus. It passes anteriorly through the middle cranial fossa in the outer aspect of the wall of the cavernous sinus. During its course to the superior orbital fissure (Figure 4.1), it is closely related to the superior cerebellar and posterior cerebral arteries.

As it enters the superior orbital fissure, it is closely related to the oculomotor and abducent nerves, and the ophthalmic and maxillary divisions of the trigeminal nerve (see chapter 5, Trigeminal Nerve). It then terminates in the muscle it supplies—the superior oblique (Figures 4.2−4.5).

Clinical Anatomy of the Cranial Nerves. DOI: http://dx.doi.org/10.1016/B978-0-12-800898-0.00004-X

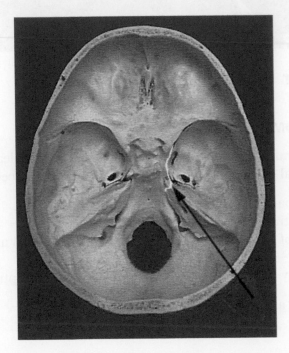

Figure 4.1 The trochlear nerve is highlighted in yellow as it passes anteriorly to enter the orbit through the superior orbital fissure. The arrow indicates the position of the trochlear nerve as it passes over the medial edge of the petrous temporal bone.

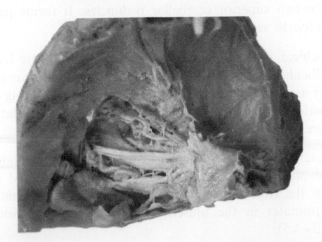

Figure 4.2 An unlabeled diagram to demonstrate the position of the trochlear nerve in relation to other structures.

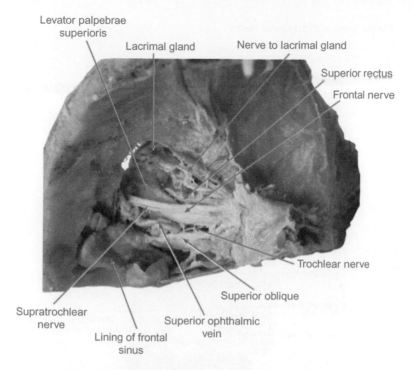

Levator palpebrae superioris

Lacrimal gland

Nerve to lacrimal gland

Superior rectus

Frontal nerve

Trochlear nerve

Superior oblique

Supratrochlear nerve

Superior ophthalmic vein

Lining of frontal sinus

Figure 4.3 A labeled diagram to demonstrate the position of the trochlear nerve in relation to other structures.

Figure 4.4 An unlabeled diagram to demonstrate the position of the trochlear nerve in relation to other structures.

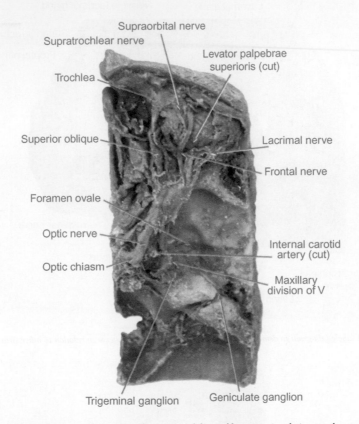

Supraorbital nerve

Supratrochlear nerve

Levator palpebrae superioris (cut)

Trochlea

Superior oblique

Lacrimal nerve

Frontal nerve

Foramen ovale

Optic nerve

Optic chiasm

Internal carotid artery (cut)

Maxillary division of V

Trigeminal ganglion   Geniculate ganglion

Figure 4.5 A labeled diagram to demonstrate the position of the trochlear nerve in relation to other structures.

The Components
Somatic Motor
The trochlear nerve has somatic motor fibers which supply a single muscle—the superior oblique. The tendon of the superior oblique passes through a pulley-like structure called the trochlea, hence the name trochlear nerve.

The Ganglia
No ganglia are associated with the trochlear nerve. The nucleus of the trochlear nerve lies deep to the fourth ventricle, beside the periaqueductal grey matter and the cerebral aqueduct, directly below the oculomotor nerve. As there is decussation at the level of origin of the trochlear nerve, a lesion of the *trochlear nucleus, or pathology before it exits the midbrain,* will affect the opposite eye.

The Important Branches

There are no known peripheral branches, although it receives connections from the corticonuclear tracts (bilaterally), medial longitudinal fasciculus (nerve pathway carrying information about the eye movements), and the tectobulbar tracts (nerve pathway coordinating head and eye movements).

THE CLINICAL APPLICATION

The first thing to remember about the superior oblique is that it has a rotational movement both in the vertical plane and also the medial/lateral planes. The superior oblique muscle initially lies behind the eye; however, its tendon passes anteriorly before reaching the superior point of the eye. Therefore, the tendon of the superior oblique has two forces—one depressing the eye and one causing rotation of the eyeball to the nose.

However, the strength of each of these components depends on the way the eye is looking. When the eye is medially rotated, depression of the eyeball is increased. When the eyeball is laterally positioned, the rotation of it towards the nose increases. When the eyeball is in the neutral position, depression and rotation towards the nose work in approximately equal amounts.

Therefore, injury to the trochlear nerve will result in weakness of the eyeball to move downward with diplopia (double vision). The patient will notice this vertical diplopia most when looking downward, e.g., when walking downstairs or reading a book.

Testing at the Bedside

Testing of the trochlear nerve is undertaken when assessing the oculomotor nerve and the abducens nerves at the same time. It involves testing of all of the extraocular muscles. Therefore, the procedure for testing the trochlear nerve is as follows:

1. Ask the patient to keep their head still during the examination
2. With their eyes only, they should follow the tip of your finger (or pen torch or pencil, etc.)
3. The examiner should then move the object in the horizontal plane from extreme left to extreme right
4. This should be done SLOWLY

5. When at the extreme left or right, with the examining object, stop!
6. Observe for nystagmus
7. Then an H-shaped pattern with the examining object can be drawn in space, which the patient should follow WITH THEIR EYES ONLY
8. All movements should be done SLOWLY to assess for eye movements.

Advanced Testing

Testing of the trochlear nerve integrity can be done at the bedside. However, as most cases of unilateral or bilateral trochlear nerve palsies are related to closed head trauma, it may be necessary to request a CT or MRI of the brain to assess the pathology in greater detail. Further investigation and intervention will depend on the clinical presentation, history, and detailed neurological examination. It is also worthwhile keeping the diagnosis of myasthenia gravis as a potential cause of weakness of the eye musculature.

PATHOLOGIES

The long intracranial course of the trochlear nerve makes it particularly vulnerable to injury from blunt head trauma or any pathologies which cause an increase in intracranial pressure. Brainstem lesions that damage the trochlear nerve nucleus are not easily identifiable clinically. The following table summarizes the main causes of a trochlear nerve palsy.

Causes of Trochlear Nerve Palsy	
Trauma	Head trauma, e.g., road traffic accident
Vascular	Arterio-venous malformation Hypertension Aneurysm Microvascular disease (e.g., due to diabetes) Intracranial hemorrhage
Neurological	Demyelinating disease, e.g., multiple sclerosis Myasthenia gravis Progressive supranuclear palsy
Infection	Meningitis Sphenoid sinusitis
Neoplastic	Any space occupying lesion within the cranial cavity
Congenital	Isolated trochlear nerve pathology
Iatrogenic	From neurosurgery

INTERESTING CLINICAL QUESTIONS

Q:

Why does diplopia classically occur when the patient looks down-wards when they have an abducent nerve pathology?

A:

Diplopia, or double vision, classically occurs when the patient looks down for two reasons. First the superior oblique normally helps with the inferior rectus in moving the eyeball downwards and is the only muscle able to do this when the eye is brought inwards (adducted). In addition to this, the superior oblique is the major muscle that results in rotation of the eyeball towards the nose. Therefore, damage, for whatever reason, to the trochlear nerve will result in inability of the superior muscle to do its two major functions: intorsion and depression of the eyeball. Therefore, the direction of gaze is different for the affected and unaffected eyeball. Therefore, when the patient looks downwards, they will experience diplopia. To try to correct this, the patient will raise their head slightly and lateral towards the unaffected eye.

Q:

What investigative tools can commonly be used to diagnose a troch-lear nerve pathology?

A:

Aside from the initial bedside investigation and full assessment of the patient's history and an eye examination, as detailed above, common investigations could include an MRI scan of the brain and blood work up to include/exclude diabetes mellitus. The presenting signs and symptoms and clinical history will direct the use of the most appropriate investigations.

CHAPTER 5

Trigeminal Nerve

THE ANATOMY—SUMMARY

The trigeminal nerve is the fifth cranial nerve. It is also the largest of all of the cranial nerves. It has three branches as follows:

1. Ophthalmic (abbreviated as CN V$_1$)—general sensory component
2. Maxillary (abbreviated as CN V$_2$)—general sensory component
3. Mandibular (abbreviated as CN V$_3$)—general sensory and branchial motor components.

The trigeminal nerve is the principal sensory nerve of the head innervating the skin of the face, mucosa of the mouth, nasal cavity, and paranasal sinuses, and most of the dura mater and the cerebral arteries.

THE ANATOMY—IN MORE DETAIL

The trigeminal nerve arises from the lateral aspect of the pons comprised of a large sensory root and a smaller motor root. Cell bodies of the trigeminal nerve are located in the trigeminal ganglion with a lesser amount in the mesencephalic trigeminal nucleus.

It is the peripheral processes of the ganglion that forms the ophthalmic and maxillary nerves and the sensory part of the mandibular nerve. In addition, within the mandibular nerve, proprioceptive fibers are present from the mesencephalic nucleus.

Central processes of the trigeminal ganglion enter the pons and then pass to the spinal and pontine trigeminal nuclei. It is the large fibers for discriminative touch that terminate in the pontine trigeminal nucleus. Indeed, the pontine trigeminal nucleus is referred to as the chief or principal sensory nucleus.

However, a smaller number of fibers pass caudally towards the spinal cord to the spinal trigeminal nucleus. The spinal trigeminal nucleus is responsible for conveying information related to light touch, pain, and temperature.

Clinical Anatomy of the Cranial Nerves. DOI: http://dx.doi.org/10.1016/B978-0-12-800898-0.00005-1

The information conveyed in the spinal trigeminal tract also includes input from the outer aspect of the ear, posterior one-third of the tongue (mucosa of), pharynx, and larynx. This means there is an input related to sensation also from these sites and is related to the facial, glossopharyngeal, and vagus nerves, respectively. In the sensory root and the spinal cord portion of the trigeminal nerve, there is a spatial arrangement of the fibers. The mandibular fibers are initially ventral and the ophthalmic fibers dorsal, with the maxillary fibers lying between these. On approach to the brainstem, there is a rotation of the fibers to lie the opposite way, i.e., the mandibular fibers end up dorsal and the ophthalmic fibers ventral, and again the maxillary fibers sandwiched between the two.

In addition to this, the mesencephalic trigeminal nucleus extends from the pontine trigeminal nucleus to the midbrain. This has two processes—a central part and a peripheral portion. The peripheral branches of the mesencephalic trigeminal nucleus pass within the mandibular nerve and terminate in proprioceptive receptors beside the teeth of the mandible and in the muscles of mastication (i.e., temporalis, masseter, and the pterygoid muscles). Occasional fibers also pass to the maxillary division ending in the hard palate adjacent to the teeth of the maxilla.

The central branches of the mesencephalic trigeminal nucleus terminate in either the motor nucleus of the trigeminal nerve or the reticular formation (and then onwards to the thalamus).

The bulk of the motor root of the trigeminal nerve contain fibers from the trigeminal motor nucleus. This nucleus, found medial to the chief or principal sensory nucleus (i.e., pontine trigeminal nucleus), supplies the muscles of mastication (temporalis, masseter, and the pterygoids (lateral and medial)). It also supplies the anterior belly of digastric, mylohyoid, tensor veli tympani, and tensor tympani. The trigeminal motor nucleus receives afferent information from the corticobulbar tract (that white matter pathway connecting the cerebral cortex to the brainstem). Afferent information also arrives from the sensory trigeminal nuclei. This pathway deals with the stretch reflex and the jaw-opening reflex.

On leaving the brainstem, the motor fibers of the trigeminal nerve pass below the ganglion along the floor of Meckel's cave (named after the German anatomist Johann Friedrich Meckel, the Elder (to avoid confusion with his famous grandson, also an anatomist)). These fibers are only present in the mandibular division of the trigeminal nerve and become related to the sensory fibers as the whole nerve passes through the foramen ovale. It goes on to supply the muscles of mastication, and some of the smaller muscles previously described.

The Components
There are two main components in the trigeminal nerve—general sensory and branchial motor.

Branchial Motor
The cell bodies of the branchial motor component arise from the pons and go on to supply the muscles of mastication, namely the temporalis, masseter, and the two pterygoids—lateral and medial. The branchial motor component also supplies the tensor veli palatini (tenses the soft palate) and the tensor tympani (tenses the tympanic membrane which reduces the noise during chewing by reducing the movement of the ossicles).

General Sensory
The cell bodies of the general sensory fibers are found within the trigeminal ganglion, found on the superior aspect of the petrous temporal bone, within Meckel's cave. The sensory fibers convey sensory information from a wide territory of the face down to the upper lip. It also conveys sensory information from the upper teeth, paranasal sinuses, nose, and cornea.

The Ganglia
There are two ganglia which are related to the trigeminal nerve—the trigeminal ganglion itself and the submandibular ganglion.

Ganglion	Location	Function
Trigeminal	Within Meckel's cave near the apex of the petrous temporal bone	Sensory ganglion
Submandibular	Above the deep portion of the submandibular gland on hyoglossus	Synapses here for pre- and postsynaptic parasympathetic innervation of the submandibular and sublingual salivary glands from the chorda tympani of the facial nerve

The Nuclei

Nucleus	Information Carried	Location	Projection
Spinal trigeminal nucleus	Pain, temperature, and light touch	Medulla	Ventral posteromedial nucleus of the dorsal thalamus
Pontine trigeminal nucleus	Discriminative and light touch, proprioception of the jaw	Pons	Ventral posteromedial nucleus of the thalamus
Mesencephalic trigeminal nucleus	Proprioception of the face (lower jaw)	Pons and midbrain	Motor nuclei of the trigeminal nerve
Trigeminal motor nucleus	Motor information	Pons	Muscles of mastication and tensor veli palatini, tensor tympani, anterior belly of digastric and mylohyoid

The Important Branches

There are three main divisions of the trigeminal nerve which will be discussed now—the ophthalmic, maxillary, and mandibular nerves.

Ophthalmic Nerve (CN V₁)

The ophthalmic division of the trigeminal nerve, also referred to as the ophthalmic nerve, is a purely sensory (afferent) nerve. It is the smallest division of the trigeminal nerve. It runs forward in the lateral wall of the cavernous sinus below the oculomotor and trochlear nerves. It divides into the frontal, lacrimal, and nasociliary nerves, which enter the orbital cavity through the superior orbital fissure (Figure 5.1). It does not supply pharyngeal arch origin structures as it is derived from the paraxial mesoderm. In general, it supplies the skin and mucous membranes of the head and nose at the front. It supplies the skin of the face above the level of the orbit, but extending down to the tip of the nose only on the anterior aspect. It does not supply the lateral aspect of the nose. These features are important in clinical testing and pathology involving a single branch of the trigeminal nerve, which will be discussed later. As well as supplying the skin of the forehead and the upper eyelid, it also supplies the cornea, paranasal sinuses, and nasal mucosa. The specific branches of the ophthalmic nerve are as follows:

1. Frontal nerve—Terminates as the supraorbital and supratrochlear nerves innervating the skin of the scalp over the area of frontalis and the frontal bone, frontal sinuses, and the upper eyelid

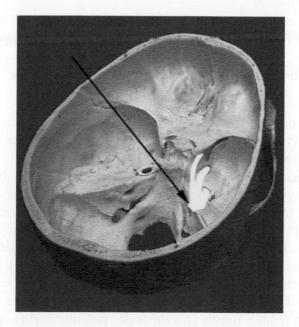

Figure 5.1 The trigeminal nerve is highlighted in yellow as it passes over the petrous temporal bone. At that point, it is covered by dura mater and referred to as Meckel's cave. It has three branches—the ophthalmic, maxillary, and mandibular divisions. The ophthalmic division passes through the superior orbital fissure, the maxillary division passes through the foramen rotundum, and the mandibular division passes through the foramen ovale.

2. Lacrimal nerve—Provides sensory innervation to the lacrimal gland, conjunctiva, and the upper eyelid
3. Nasociliary nerve—Supplies the nasal mucous membrane, paranasal sinuses and is also involved in afferent loop of the corneal blink reflex
4. Meningeal branch (tentorial nerve)—Arising from the intracranial part of the ophthalmic nerve. This branch supplies the falx cerebri (supratentorial portion) and the tentorium cerebelli.

Maxillary Nerve (CN V$_2$)

The maxillary division of the trigeminal nerve, also referred to as the maxillary nerve, is a purely sensory (afferent) nerve. It is the medium-sized branch of the trigeminal nerve between the smaller ophthalmic division and the largest mandibular division. After emerging from the trigeminal ganglion, it passes to the pterygopalatine fossa, passing to the posterior surface of the maxilla before passing through the foramen rotundum and entering the orbit through the inferior orbital fissure, running here to terminate on the anterior aspect of the skull at the infraorbital foramen (Figure 5.1). The maxillary nerve also passes through the cavernous sinus. In general, it supplies the teeth of the maxilla, skin from the lower eyelid above to the superior aspect of the mouth below, as well as the nasal cavity and the paranasal sinuses. The specific branches of the maxillary nerve are as follows:

1. Middle meningeal branch—To supply the dura
2. Alveolar nerves (anterior, middle, and posterior superior alveolar nerves)—Provide sensory innervation to all of the upper teeth in the maxilla as well as the gingiva
3. Zygomatic nerve—Supplying the skin of the side of the forehead (via the zygomaticotemporal nerve) and the area over the prominence of the cheek (maxilla) anteriorly (via the zygomaticofacial nerve). This nerve also carries with it parasympathetic postganglionic fibers from the facial nerve to innervate the lacrimal gland
4. Palatine nerves (greater and lesser palatine nerves, as well as the nasopalatine nerve)—To supply the gingiva, mucous membranes of the roof of the mouth (via the greater palatine nerve), soft palate (including uvula), tonsils (via the lesser palatine nerve), and the palatal structures around the superior anterior six teeth (via the nasopalatine nerve)
5. Pharyngeal nerve—A small branch which passes to the area behind the auditory tube supplying the nasopharynx for its mucosa
6. Infraorbital nerve—Exits through the infraorbital foramen of the maxilla to supply the lower eyelid and superior lip
7. Inferior palpebral nerve—Supplies the skin of the inferior eyelid as well as the conjunctiva
8. Superior labial nerve—Supplies the skin of the superior lip and the mucosa of the mouth at this point
9. External nasal branches.

Mandibular Nerve (CN V₃)

The mandibular division of the trigeminal nerve, also referred to as the mandibular nerve, is a mixed sensory and branchial motor nerve. It is also the largest of the three branches of the trigeminal nerve. The sensory root arises from the lateral aspect of the ganglion, with the motor division lying deeper. In general, the mandibular nerve supplies the lower face for sensation over the mandible, including the attached teeth, the temporomandibular joint (TMJ), and the mucous membrane of the mouth as well as the anterior two-thirds of the tongue (the posterior one-third is supplied by the glossopharyngeal nerve). It also supplies the muscles of mastication which are the medial and lateral pterygoids, temporalis, and masseter. It also supplies some smaller muscles namely the tensor veli tympani, tensor veli palatini, mylohyoid, and the anterior belly of digastric.

The mandibular nerve enters the infratemporal fossa, passes through the foramen ovale in the sphenoid bone, and divides at that point into a smaller anterior and a larger posterior trunk (Figure 5.1). The main trunk gives off two branches at this point. The first is a meningeal branch and passes through the foramen spinosum to receive innervation from the meninges of the middle cranial fossa. The second small branch, a muscular branch, which supplies the medial pterygoid and also a twig to the otic ganglion to supply the tensor veli palatini and the tensor tympani. Two main divisions arise from the main trunk of the mandibular nerve after these two smaller branches have been given off: an anterior and posterior division (Figures 5.2—5.5).

Anterior Division
1. Masseteric nerve—Passing posterior to the tendon of temporalis, this branch approaches the masseteric muscle on its deep aspect

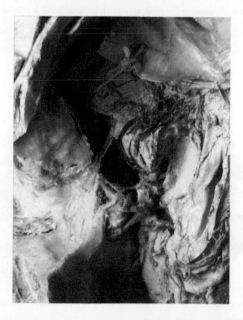

Figure 5.2 An unlabeled image to demonstrate the position of the trigeminal nerve in relation to other structures.

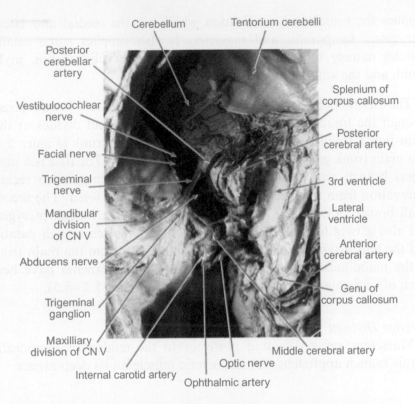

Cerebellum   Tentorium cerebelli

Posterior
cerebellar
artery

Vestibulocochlear
nerve

Facial nerve

Trigeminal
nerve

Mandibular
division
of CN V

Abducens nerve

Trigeminal
ganglion

Maxilliary
division of CN V

Internal carotid artery

Optic nerve

Ophthalmic artery

Splenium of
corpus callosum

Posterior
cerebral artery

3rd ventricle

Lateral
ventricle

Anterior
cerebral artery

Genu of
corpus callosum

Middle cerebral artery

Figure 5.3 A labeled image to demonstrate the position of the trigeminal nerve in relation to other structures.

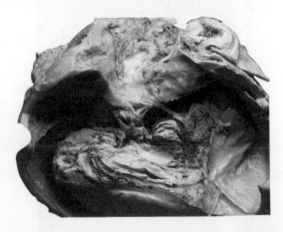

Figure 5.4 An unlabeled image to demonstrate the position of the trigeminal nerve in relation to other structures.

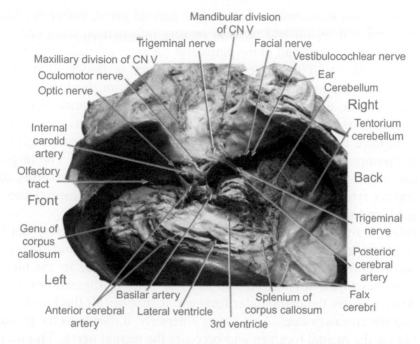

Figure 5.5 A labeled image to demonstrate the position of the trigeminal nerve in relation to other structures.

2. Deep temporal nerves—Two branches arise generally from this—an anterior and posterior division. Sometimes, a third (intermediate) branch may be found

3. Lateral pterygoid nerve—This branch enters the deep surface of the muscle

4. Buccal branches— TIP: DO NOT CONFUSE THIS WITH THE BUCCAL BRANCH OF THE FACIAL NERVE WHICH CONVEYS MOTOR INFORMATION TO BUCCINATOR. This buccal branch from the anterior division of the mandibular nerve passes between the lateral pterygoid heads then inferior to the temporalis tendon passing to the buccal membrane to receive sensory information from that site, skin over the cheek, as well as the second and third molars.

Posterior Division

1. Auriculotemporal nerve—This branch has two roots and is closely related to the middle meningeal artery. It passes postero-superiorly behind the TMJ. It is closely related to the superficial temporal vessels

and gives off secretomotor fibers to the parotid gland, before reaching the temporal region and receives sensory information from here, as well as the superior half of the pinna and the external auditory meatus.

2. Lingual nerve—This nerve carries two types of information—that related to the trigeminal nerve for sensation from the tongue (general somatic afferent) but also carries with it a branch from the facial nerve—the chorda tympani nerve for special sensory fibers of taste from the front two-thirds of the tongue, but also preganglionic parasympathetic fibers to the submandibular ganglion. The lingual nerve first passes below the lateral pterygoid muscle, receives the chorda tympani nerve and then passes between the medial pterygoid, and then passes towards the tongue.

3. Inferior alveolar nerve—This is a dual motor and sensory nerve. It passes on the medial aspect of the lateral pterygoid and just before entering the mandibular foramen, it gives rise to the motor branches to the mylohyoid and anterior belly of digastric muscles. After it enters the mandibular foramen, it supplies all the lower teeth and the alveolar ridges. As it passes anteriorly, it then exits the mandible via the mental foramen and becomes the mental nerve. This nerve supplies sensation to the skin over the chin. This nerve is crucial in dental practice, as it is the nerve that is anesthetized as it enters the mandibular foramen to provide complete nerve block if procedures are to be undertaken on the lower teeth or related structures.

All the branches of the trigeminal nerve and structures it supplies can be seen in Figures 5.2–5.5.

THE CLINICAL APPLICATION

Testing at the Bedside

There are two aspects that can be tested for clinically with the trigeminal nerve—the motor function of the nerve and if sensation is intact.

1. Ensure you take a detailed clinical history first.
2. ALWAYS tell the patient what you will be doing and what you expect them to do in helping elicit any signs and/or symptoms.
3. Observe the skin over the area of temporalis and masseter first to identify if any atrophy or hypertrophy is obvious.

4. First, palpate the masseter muscles while you instruct the patient to bite down hard. Also note masseter wasting on observation. Do the same with the temporalis muscle.

5. Then, ask the patient to open their mouth against resistance applied by the instructor at the base of the patient's chin.

6. To assess the stretch reflex (jaw jerk reflex), ask the patient to have their mouth half open and half closed. Place an index finger onto the tip of the mandible at the mental protuberance, and tap your finger briskly with a tendon hammer. Normally this reflex is absent or very light. However, for patients with an upper motor neuron lesion, the stretch reflex (jaw jerk reflex) will be exaggerated.

7. Also ask the patient to move their jaw from side to side.

8. Next, test gross sensation of the trigeminal nerve. Tell the patient to close their eyes and say "sharp" or "dull" when they feel an object touch their face. Allowing them to see the needle, brush, or cotton wool ball before this examination may alleviate any fear. Using the needle, brush, or cotton wool randomly touch the patient's face with the object. Touch the patient above each temple, next to the nose and on each side of the chin, all bilaterally. You must test each of the territories of distribution of the ophthalmic, maxillary, and mandibular nerves.

9. Ask the patient to also compare the strength of the sensation of both sides. If the patient has difficulty distinguishing pinprick and light touch, then proceed to check temperature and vibration sensation using the vibration fork. You can heat it up or cool it down in warm or cold water, respectively.

10. Finally, test the corneal reflex (blink reflex). You can test it with a cotton wool ball rolled to a fine tip. Ask the patient to look at a distant object and then approaching laterally, touch the cornea (and not the sclera) looking for the eyes to blink. Repeat this on the opposite eye. If there is possible facial nerve pathology on the side that you are examining, it is imperative to observe the opposite side for the corneal reflex.

11. Some clinicians omit the corneal reflex unless there is sensory loss on the face elicited from the history or examination, or if cranial nerve palsies are present at the pontine level. It is best to ensure a complete clinical examination is undertaken, however, especially if there is a possible pathology of the trigeminal nerve.

Tip!

When examining the sensory distribution of the trigeminal nerve, do not test each territory of nerve supply consecutively, i.e., do not do left and right ophthalmic, left and right maxillary, and left and right mandibular nerve territories. The patient will start to predict what you will do next so it is best to do it at random. Just ensure you examine all territories, both left and right sides.

Advanced Testing

Nerve conduction studies and detailed muscular studies may be used, but a lot can be determined from the simple examinations detailed above. It may be necessary to have an MRI scan to further investigate any intracranial pathology, which is the imaging modality preferred (Woolfall and Coulthard, 2001). Further specialist input may be necessary from a neurologist.

PATHOLOGIES

Trigeminal Neuralgia

Trigeminal neuralgia, or tic douloureux (Fr. painful twitch), has a prevalence of 0.1–0.2 per 1000 and an incidence ranging from 4–5 per 100,000/year up to approximately 20 per 100,000/year after the age of 60 (Manzoni and Torelli, 2005). It affects twice as many women as men and is rare below the age of 40 (NHS[1]). In this condition, it causes excruciating pain in one of the territories of distribution of the trigeminal nerve, with the maxillary nerve the most commonly affected, followed by the mandibular nerve and then less commonly, the ophthalmic nerve. Generally, it affects only one side of the face and the most common cause is pressure on the trigeminal nerve within the skull.

The condition presents as severe facial pain described by patients as "stabbing, shooting, excruciating, or burning" (Trigeminal Neuralgia Association UK[2]). Touching the affected area, even by light touch, often sets off the facial pain. It can last less than a second up to a couple of minutes, and there may be pain-free periods for months or years. It can be a very debilitating condition as day-to-day activities can

---

[1]http://www.nhs.uk/conditions/trigeminal-neuralgia/Pages/Introduction.aspx (accessed 07.01.14).
[2]http://www.tna.org.uk (accessed 08.01.14).

bring on the pain, e.g., the wind blowing on the face, chewing, speaking, eating and drinking, and shaving.

There are three different types of trigeminal neuralgia, types 1, 2 and symptomatic trigeminal neuralgia referred to as TN1, TN2, and STN, respectively. TN1 is the classical form where the pain only occurs occasionally and is not constant. There is no identifiable cause for TN1. TN2 is referred to as atypical trigeminal neuralgia as the pain is more constant and is like a throbbing sensation. STN is where there is an identifiable cause, e.g., multiple sclerosis.

In most patients, the antiepileptic agent carbamazepine can be given. Although it is an antiepileptic agent, it works by reducing nerve impulses, hence dulls or eliminates the pain of the acute attack. If medical treatment does not work, it may be necessary to consider surgery for those patients severely affected by trigeminal neuralgia, yet unresponsive to medication.

Potential procedures which can be undertaken if medication is ineffective are thermocoagulation of the nerve endings, balloon compression, electric currents applied to the branch of the trigeminal nerve, peripheral radiofrequency, or glycerol injection of the trigeminal nerve branch.

Many cases of trigeminal neuralgia with no systemic cause are due to the presence of a small aberrant artery pressing against a branch of the trigeminal nerve. Microvascular decompression surgery has been shown to be highly successful with over 70% of individuals pain free after 10 years postoperatively (NHS[1]). One newer procedure is a stereotactic procedure referred to as gamma knife radiosurgery. Here, a concentrated beam of radiation is directed towards the trigeminal nerve. It is a controversial area with emerging research demonstrating that patients treated multiple times with this surgery are more likely to get facial numbness than those who have had it a single time (Elaimy et al., 2012). Questions still remain about the efficacy of this procedure.

Herpes Zoster

The varicella zoster virus causes herpes zoster (shingles), the most common sensory abnormality of the face and scalp. It manifests in childhood as chickenpox (vesicular skin rash) but the virus is never eliminated from the body. Following the initial acute illness, it then remains dormant in the nerve cell bodies, including the trigeminal ganglion. Years or decades after the initial episode of chicken pox, the virus "wakens up" and passes down the nerve axons to cause a viral infection of the skin again. This process can take a few days and there is no obvious reason for it to flare up though most patients are over 50 years old that get herpes zoster. If a patient under 50 has herpes zoster or it affects more than one dermatome, an immunodeficiency should be suspected.

It results in a burning pain, itching (which can be severe), and a vesicular skin eruption. It affects one of the divisions of the trigeminal nerve, i.e., ophthalmic, maxillary, or mandibular nerves. It can be a very serious condition if the ophthalmic division is affected as it can result in corneal ulceration, which can threaten the integrity of the eye, and therefore sight. Therefore, if ophthalmic herpes zoster is suspected, an ophthalmologist must be consulted immediately.

In terms of treating herpes zoster, topical antiviral treatment has been shown to be ineffective. Oral acyclovir has been shown to reduce the duration of the signs and symptoms, and reduce the rate and severity of complications related to ophthalmic herpes zoster.

The most common complication from herpes zoster is postherpetic neuralgia (PHN). Other complications of herpes zoster relate to ocular, and surrounding structures, therefore input from an ophthalmologist is essential.

TMJ Dysfunction

Temporomandibular disorders (TMDs) are a collection of conditions affecting the TMJ. It is a relatively common condition affecting approximately 12% of the population, but with many factors that contribute to it (Marklund and Wänman, 2007).

The main symptoms of TMJ disorders are facial pain (around the TMJ but can be referred to the ear, head, and neck), restricted jaw movement (in any plane, with movement increasing it), and noise from movement of the TMJ (e.g., clicking on movement). Other symptoms can include headache, ear pain, dizziness, tinnitus, or locking of the jaw.

The main problems causing TMJ disorders are degenerative change (e.g., osteoarthritis and rheumatoid arthritis), intra-articular disc abnormality, gout, trauma, or hypermobility (and also hypomobility).

Treatment of TMD depends on the initial cause. If there is an obvious anatomical abnormality amenable to surgery, then this will help to resolve the problem. However, explanation and reassurance can be used as most cases of TMD are benign and improve through time. Other therapies can include intra-articular steroid injections, acupuncture, and bite guards (occlusal splints). Occasionally, analgesics or antidepressants may need to be prescribed.

INTERESTING CLINICAL QUESTIONS

Q:
If the trigeminal nerve was injured, what will be able to be elicited on the clinical examination?

A:
There are a wide variety of pathologies, which can result in damage to the trigeminal nerve as described earlier. Obviously, it depends if the entire trigeminal nerve has been damaged or one of its specific branches namely the ophthalmic, maxillary, or mandibular nerves. The patient may have complete or partial paralysis of the muscles of mastication. On examination, the mandible will be deviated towards the side of the lesion. There will be a loss of sensation to the face for light touch, temperature, or painful sensations. In addition, there will be no corneal reflex.

Q:
What is an inferior alveolar nerve block?

A:
This is one of the most common dental procedures undertaken. It is where the inferior alveolar nerve is anesthetized just before it enters

the mandible at the mandibular foramen. When the anesthetic agent takes effect, it results in anesthesia of the lower teeth and related gingivae, and also the skin of the chin and the lower lip, as the inferior alveolar nerve terminates as the mental nerve as it exits the mental foramen.

Q:
Following a nerve block of the inferior alveolar nerve for dental procedures of the lower teeth and surrounding structures, why can loss of sensation to the tongue happen?

A:
The lingual nerve is found very close to where the inferior alveolar nerve is anesthetized as it enters the mandibular foramen. As the lingual nerve supplies sensation to the anterior two-thirds of the tongue, it can also be anesthetized resulting in a temporary loss of sensation to that portion of the tongue.

Q:
The inferior alveolar nerve is readily accessible for dental anesthesia; however, the nerves supplying the maxillary teeth are not as readily accessible. How is an anesthetic agent administered to the maxillary teeth?

A:
The anesthetic agent is administered into the soft tissue surrounding the roots of the teeth. A period of time must elapse to allow the anesthetic agent to infiltrate the nerve roots of the superior alveolar nerves.

REFERENCES

Elaimy, A.L., Hanson, P.W., Lamoreaux, W.T., Mackay, A.R., Demakas, J.J., Fairbanks, R.K., et al., 2012. Clinical outcomes of gamma knife radiosurgery in the treatment of patients with trigeminal neuralgia. Int. J. Otolaryngol. 2012, 13. Article ID 919186, doi:10.1155/2012/919186.

Manzoni, G.C., Torelli, P., 2005. Epidemiology of typical and atypical craniofacial neuralgias. Neurol. Sci. 26 (Suppl. 2), s65—s67.

Marklund, S., Wänman, A., 2007. Incidence and prevalence of temporomandibular joint pain and dysfunction. A one-year prospective study of university students. Acta Odontol. Scand. 65, 119—127.

Woolfall, P., Coulthard, A., 2001. Trigeminal nerve: anatomy and pathology. Br. J. Radiol. 74, 458—467.

CHAPTER 6

Abducent Nerve

THE ANATOMY—SUMMARY

The abducent nerve is the sixth cranial nerve. It is a somatic motor nerve or a general somatic efferent nerve with proprioceptive fibers present. It supplies only one muscle—the extraocular lateral rectus muscle.

THE ANATOMY—IN MORE DETAIL

The abducent nerve nucleus is found in the pons at the level of the facial colliculus, at the floor of the fourth ventricle. The facial nerve axons loop round its nucleus creating a bulge into the fourth ventricle—the facial colliculus. Like the other nuclei for the extraocular muscles, i.e., the oculomotor and trochlear nuclei, the abducent nucleus is also found close to the midline. The motor axons in this nerve then pass ventral and caudal passing to the corticospinal (contains primarily motor fibers) before leaving at the pontomedullary junction.

From here, it runs superior, anterior, and lateral through the pontine cistern (space on the ventral part of the pons) before passing dorsal to the anterior inferior cerebellar artery. It then passes into the dura inferior to the posterior clinoid process crossing the inferior petrosal sinus. At this point, it has the longest intradural course of any of the cranial nerves. It then passes over the superior border of the apical region of the petrous temporal bone bending sharply anteriorly. From here, it then passes anterior to run in the cavernous sinus, lateral to but parallel with, the internal carotid artery. It enters the superior orbital fissure in the common tendinous ring, entering the lateral rectus on its medial side (Figure 6.1).

The Components

The abducent nerve contains general somatic efferent and proprioceptive fibers destined for a single muscle—the lateral rectus. In addition, postganglionic sympathetic fibers are also present (Figures 6.2—6.5).

Clinical Anatomy of the Cranial Nerves. DOI: http://dx.doi.org/10.1016/B978-0-12-800898-0.00006-3

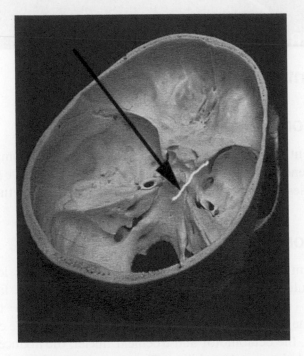

Figure 6.1 The abducent nerve is highlighted in yellow as it passes into the superior orbital fissure.

Figure 6.2 An unlabeled image to demonstrate the position of the abducent nerve in relation to other structures.

The Ganglia
There are no ganglia associated with the abducent nerve.

The Important Branches
There are no established branches that arise from the abducent nerve.

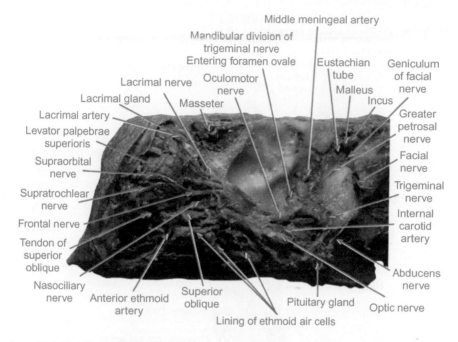

Figure 6.3 A labeled image to demonstrate the position of the abducent nerve in relation to other structures.

Figure 6.4 An unlabeled image to demonstrate the position of the abducent nerve in relation to other structures.

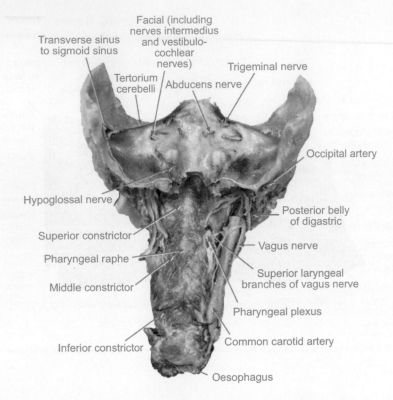

Figure 6.5 A labeled image to demonstrate the position of the abducent nerve in relation to other structures.

THE CLINICAL APPLICATION

Testing at the Bedside

Testing of the abducent nerve is undertaken when assessing the oculomotor nerve and the trochlear nerves at the same time. It involves testing of all of the extraocular muscles. Therefore, the procedure for testing the abducent nerve is exactly the same as that for the oculomotor nerve and the trochlear nerves:

1. Ask the patient to keep their head still during the examination
2. With their eyes only, they should follow the tip of your finger (or pen torch, or pencil, etc.)
3. The examiner should then move the object in the horizontal plane from extreme left to extreme right.
4. This should be done SLOWLY
5. When at the extreme left or right, with the examining object, stop!
6. Observe for nystagmus

7. Then an H-shaped pattern with the examining object can be drawn in space, which the patient should follow WITH THEIR EYES ONLY
8. All movements should be done SLOWLY to assess for eye movements.

Advanced Testing

Advanced testing can be undertaken by a variety of means, generally with the input of a specialist, e.g., ophthalmologist and/or neurological input. Three broad categories of detailed examination can be done, including the rotational movement of each eye (as previously described), comparison of yoke muscles (pairs of muscles that move the eyes in conjugate direction), and the red lens diplopia test (using diplopia as a test for weakness of the eye muscles) (Walker et al., 1990). These advanced tests, used with specialist input, can help establish pathology of all of the extraocular muscles and the nerves that supply them, i.e., oculomotor, trochlear, and abducent nerves.

In addition, further testing will be directed by what is found on clinical examination. As the abducent nerve has a long course from its origin at the pontomedullary junction to its termination in the lateral rectus, there are many causes for compression of this nerve. Any cause of an increased intracranial pressure, aneurysm, space occupying lesion, cavernous sinus thrombosis, or atherosclerotic plaque within the internal carotid artery may give rise to pressure on the abducent nerve. Therefore, the presentation of the patient, detailed history, and clinical examination will direct the most appropriate investigations, in consultation with the relevant specialist.

PATHOLOGIES

Pathologies of the abducent nerve can be divided into those affecting the nucleus (and above) and the peripheral component of the nerve.

Lesion of the Abducent Nerve Nucleus

The abducent nerve nucleus contains two types of fibers in it: those that control the ipsilateral abducent nerve and interneurons that connect with the oculomotor nucleus of the opposite side, by crossing the midline. This allows for controlled eye movement when lateral gaze of one eye is coupled with medial gaze of the contralateral eye (by the medial rectus of that side). This conjugate gaze is controlled by the medial longitudinal fasciculus. Lesions of this pathway and the abducent nerve result in *internuclear ophthalmoplegia*. When you ask the patient to look to the unaffected eye, the affected eye shows only minimal adduction. On inspection of the contralateral eye (to the site of pathology), that eye will abduct but will have nystagmus. The patient will complain of diplopia (double vision) on looking towards the *unaffected* eye. In younger patients with multiple sclerosis, bilateral internuclear ophthalmoplegia can be found.

There is a rare syndrome referred to as Wall-Eyed Bilateral INternuclear Ophthalmoplegia (WEBINO syndrome). Here the patient will have a bilateral exotropia on primary gaze, bilateral internuclear ophthalmoplegia with impaired convergence (Chakravarthi et al., 2013). Many causes have been identified for this pattern, but the most common is infarction at the level of the midbrain (Chen and Lin, 2007).

If the lesion affects the abducent nerve nucleus or the paramedian pontine reticular formation, as well as the medial longitudinal fasciculus on the same side, it will result in conjugate horizontal gaze palsy in one direction and internuclear ophthalmoplegia in the other. This is referred to as the one and a half syndrome, typically caused by multiple sclerosis, brainstem stroke or tumor, or arteriovenous malformations at the level of the brainstem (Wall and Wray, 1983).

Lesions of the Peripheral Abducent Nerve

A variety of pathologies can cause an abducent nerve paralysis. The following table highlights some of the more common reasons for this.

Causes of Abducent Nerve Palsy	
Trauma	Head trauma, e.g., base of skull fractures
Vascular	Aneurysm Cavernous sinus disease Infarction (stroke) Intracerebral hemorrhage
Neurological	Diabetic neuropathy (most common cause) Variety of neuropathies Demyelination, e.g., multiple sclerosis
Infection	Meningitis
Neoplastic	Any space occupying lesion within the cranial cavity compressing on the abducent nerve
Congenital	An isolated abducent nerve pathology in a child should always be further investigated for an intracranial tumor as the cause
Iatrogenic	The abducent nerve is the most common nerve to be injured from insertion of a halo orthosis for cervical spinal injuries (Benzel, 2012)

INTERESTING CLINICAL QUESTIONS

Q:

What would be found on clinical examination of a patient with a pathology compressing the peripheral part of the abducent nerve?

A:

The affected eye would be adducted and on examination would not be able to be brought out further than the midline. Therefore, abduction of the affected eye would not be present.

Q:

How would a patient try to compensate for the diplopia experienced with an abducent nerve palsy?

A:

The patient typically may have diplopia on tasks that require the head to be relatively central but with peripheral vision to be used by rotation of the eyes, e.g., as in driving. To compensate for the diplopia, the patient will rotate their head in the direction of the side of the pathology to allow both eyes to look over to the affected side. This minimizes the diplopia when looking towards the affected side.

REFERENCES

Benzel, E.C., 2012. The Cervical Spine, fifth ed. Lippincott Williams and Wilkins, Philadelphia, PA.

Chakravarthi, S., Kesav, P., Khurana, D., 2013. Wall-eyed bilateral internuclear ophthalmoplegia with vertical gaze palsy. QJM. doi:10.1093/qjmed/hct021. Available from: <http://qjmed.oxford journals.org/content/early/2013/01/25/qjmed.hct021.full> (accessed 09.01.14.)

Chen, C.M., Lin, S.H., 2007. Wall-eyed bilateral internuclear ophthalmoplegia from lesions at different levels in the brainstem. J. Neuroophthalmol. 27, 9–15.

Walker, H.K., Hall, W.D., Hurst, J.W., 1990. Clinical methods, The History, Physical, and Laboratory Examinations, third ed. Butterworths, Boston, ISBN-10: 0-409-90077-X. Available from: <http://www.ncbi.nlm.nih.gov/books/NBK201/>. (accessed 09.01.14.)

Wall, M., Wray, S., 1983. The one-and-a-half syndrome: a unilateral disorder of the pontine tegmentum—a study of 20 cases and review of the literature. Neurology 33, 971–978.

CHAPTER 7

Facial Nerve

THE ANATOMY—SUMMARY

The seventh cranial nerve is the facial nerve. It has two parts:

1. *Larger component*
   a. *motor* fibers
   b. supplies the *muscles of facial expression* (most important clinically!).
2. *Smaller component*
   a. the *nervus intermedius* (or intermediate nerve as it is sandwiched between the facial and vestibulocochlear nerves)
   b. *sensory* and *parasympathetic* fibers
   c. supplies the *anterior two-thirds of the tongue* (taste fibers) and *lacrimal, submandibular, and sublingual salivary glands* (secretomotor).

THE ANATOMY—IN MORE DETAIL

It has a short course within the cranial cavity after emerging from the junction between the pons and medulla just lateral to the root of the sixth nerve. The facial nerve (motor root and nervus intermedius), accompanied by the eighth cranial nerve (the vestibulocochlear nerve), enters the internal auditory meatus, traveling in a lateral direction through the petrous temporal bone (Figure 7.1). At the point it meets the cavity of the middle ear, it turns backward sharply forming a "knee-shaped bend." This is also where the sensory ganglion, the geniculate (GENU, L. knee) ganglion, is found. On reaching the posterior wall of the middle ear, it then passes inferiorly to exit the skull at the stylomastoid foramen. The facial nerve then enters the parotid gland, giving rise to its terminal branches for the facial muscles.

The Components

The facial nerve is composed of several different components that allow it to carry out its various functions. The divisions of the facial nerve functions are broken down into branchial (arising from the branchial/pharyngeal arches) motor, parasympathetic, and sensory (general *and* special) fibers. An overview of what each division supplies is

Clinical Anatomy of the Cranial Nerves. DOI: http://dx.doi.org/10.1016/B978-0-12-800898-0.00007-5

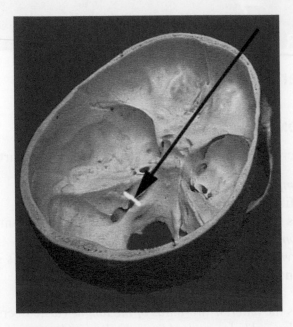

Figure 7.1 The facial nerve is highlighted in yellow as it passes into the internal acoustic (auditory) meatus of the petrous temporal bone.

detailed here. The specific branches that contain each of these types of fibers are discussed later.

Branchial Motor

The branchial motor component of the facial nerve provides innervation of the following muscles:

- Facial and auricular muscles (derived from the mesoderm (middle of the three primary germ cell layers))
- Posterior belly of digastric, stylohyoid, and stapedius
- The muscles of facial expression arise from the second pharyngeal/branchial (hyoid) arch.

Presynaptic Parasympathetic (Visceral Motor)

The presynaptic parasympathetic fibers (visceral motor) innervate three glands—two salivary glands and the lacrimal gland in the orbit:

- Lacrimal gland
- Submandibular and sublingual salivary glands
- Parasympathetic fibers synapse in ganglia (see Ganglia section) supplying each of these glands

- The lacrimal gland receives its parasympathetic innervation via the pterygopalatine ganglion
- The submandibular and sublingual salivary glands receive their parasympathetic innervation via the submandibular ganglion.

Special Sensory
The facial nerve carries with it taste fibers from the anterior two-thirds of the tongue. Specifically, this is carried in the *chorda tympani* branch which joins with the *lingual nerve* from the *mandibular division of the trigeminal nerve.*

General Sensory
From the geniculate ganglion, where the primary sensory cell bodies are located, general sensory information is relayed in the facial nerve from a small portion of the concha of the auricle.

The Ganglia
The following table summarizes the ganglia associated with the facial nerve, including its location and function(s).

Ganglion	Location	Function
Geniculate	Facial canal (petrous temporal bone)	Sensory ganglion Special sensory neuronal cell bodies for taste Fibers from motor, sensory, and parasympathetic functions pass through here
Pterygopalatine	Pterygopalatine fossa	Synapses here for pre- and postsynaptic parasympathetic innervation of the lacrimal glands Sensory and sympathetic fibers also pass through it
Submandibular	Above the deep portion of the submandibular gland on hyoglossus	Synapses here for pre- and postsynaptic parasympathetic innervation of the submandibular and sublingual salivary glands

The Important Branches
Intratemporal
Within the petrous temporal bone, the following are the major branches relevant to clinical practice:

- **Greater petrosal nerve**
  This branch arises from the geniculate ganglion and is joined by the nerve of the pterygoid canal. It contains secretomotor fibers for the lacrimal and nasal glands.

- **Nerve to stapedius**
  Supplies the stapedius muscle, which is responsible for "dampening down" loud noises protecting the middle and inner ear structures.
- **Chorda tympani nerve**
  This nerve joins the lingual nerve (of the mandibular division of the trigeminal nerve—see Chapter 5) and is distributed to the anterior two-thirds of the tongue. It contains the following:
  - Taste and sensation fibers from the front two-thirds of the tongue and the soft palate
  - Preganglionic secretory and vasodilator fibers that synapse in the submandibular ganglion.

Extratemporal

After leaving the stylomastoid foramen, the facial nerve then enters the parotid gland and starts dividing at the pes anserinus (goose's foot). The major branches are detailed including the structures they supply on the face.

- Temporal—frontalis, muscles of the external ear
- Zygomatic—remainder of frontalis, two parts of orbicularis oculi and adjacent muscles
- Buccal—upper half of orbicularis oris, buccinator, and dilator muscles inserting into upper lip
- Marginal mandibular—muscles of the lower lip
- Cervical—platysma.

Other branches of the extratemporal portion of the facial nerve include the posterior auricular (supplying posterior auricular muscles) and branches to the posterior belly of digastric and the stylohyoid.

THE CLINICAL APPLICATION

The facial nerve is tested chiefly in regard to the facial muscles as follows:

Testing at the Bedside
1. The purpose of the facial nerve is to ensure functioning of the muscles of facial expression
2. Inspect the face during conversation and rest noting any asymmetry (e.g., drooping, sagging, and even smoothing of the normal facial creases)

3. Then do the following:
   a. Ask the patient to raise their eyebrows, then
   b. Ask the patient to frown, then
   c. Ask the patient to show you their teeth

Tip!

Don't ask them to "smile" as they may be very worried about their signs and symptoms/clinical condition, and may feel uncomfortable being asked to smile! On the other hand, make sure they have their own teeth/ substitutes in place to prevent embarrassment.

   d. Puff out their cheeks against pursed lips
   e. Scrunch up their eyes, and as the examiner, try to open them on behalf of the patient

Tip!

Do tell them what you are about to do, as the patient will feel surprised that you are trying to prise their eyes open!

4. The purpose of the examination is to note asymmetry and also to determine the strength (or weakness) of the power of the facial muscles.

Tip!

Always test the resistance of the muscles of facial expression of the patient, as the patient may try to disguise their problem as they may be in denial of their underlying condition.

Advanced Testing
Taste can be tested on the anterior two-thirds of the tongue with a swab dipped in a flavored solution to test sweet/salt/sour and/or bitter substances on one half of the protruded tongue, on the side to be formally examined. Lacrimation is rarely tested.

Tip!

Always get the patient to rinse their mouth out after testing each of the standard tastes. Typically, this involves using sugar or salt, respectively.

The patient should be told what you will do before doing it. A small tongue blade can be used to place a small amount of salt or sugar onto each part of the tongue needing testing. It should be placed on the lateral aspect of the tongue. Remember when testing the facial nerve for salt or sugar taste sensation, it only supplies the anterior two-thirds.

The corneal reflex may be performed where a blink reflex is initiated in the patient by *gently* touching the corneal surface with a soft object, e.g., cotton wool. The afferent (going to the brain) arc is with the trigeminal nerve (ophthalmic division, see Chapter 5). The efferent component is via the facial nerve, so this test actually tests two of the cranial nerves at the same time—five (trigeminal) and seven (facial).

PATHOLOGIES

Facial paralysis is not an uncommon clinical condition with numerous etiologies. From a clinical perspective, it is most important to differentiate between an upper motor neuron pathology and a lower motor neuron pathology.

Upper Motor Neuron Pathology

The commonest cause of an upper motor neuron pathology is a stroke (cerebrovascular disease), though other causes can include intracranial tumors, infections (e.g., HIV and syphilis), or vasculitic diseases.

A stroke results in interruption of the fibers from the motor regions of the cerebral cortex as they pass through the internal capsule to the facial motor nucleus. This causes a voluntary paralysis of the muscles of facial expression by the facial nerve on the opposite side of the lesion. However, the upper part of the facial motor nucleus receives both crossed and uncrossed fibers resulting in the orbicularis oculi and frontalis being spared to varying degrees.

Key point: If a patient has had a stroke involving the supranuclear fibers to the facial nucleus, they are still able to wrinkle their forehead and close the eye on the contralateral side. This is despite the marked paralysis of the rest of the face, including a droopy corner of the mouth and puffing of the cheek.

Lower Motor Neuron Pathology

In a lower motor neuron paralysis involving the facial nerve, all muscles on the side of the pathology are affected, including frontalis and the orbicularis oculi. Therefore, these patients are unable to wrinkle their forehead or close their eye on the affected side. The most common cause for a lower motor neuron pathology involving the facial nerve is Bell's palsy, named after the famous Scottish anatomist and surgeon Sir Charles Bell (1774–1842) who first described it. The other causes of a lower motor neuron pathology involving the facial nerve are summarized below.

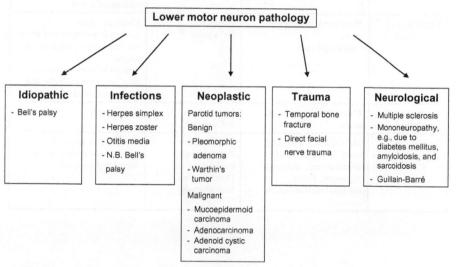

Today, Bell's palsy, the commonest cause of a facial paralysis, is understood to involve a viral infection (many have been implicated). The infection then results in inflammation and compression within the facial canal. As this canal is very small, and no room exists for expansion, the facial nerve is compressed due to edema of the tissues lining the canal pressing on the facial nerve. This reduces or blocks the nervous impulse transmission to the muscles of facial expression resulting in weakness or paralysis of those muscles.

Dependent on the site of pathology determines the clinical presentation of the patient. A summary table of what the patient may present with, taking into account the anatomical components of the facial nerve that may be affected, is given below.

Site of Pathology	Branches Affected	Type of Pathology	Clinical Presentation
Proximal to the geniculate ganglion	Main trunk of the facial nerve Nerve to stapedius Chorda tympani	Acoustic neuroma (vestibular Schwannoma) Other intracranial tumor	All functions of the nerve affected: Facial paralysis Hyperacusis Impaired secretion of tears Loss or impairment of taste on the anterior two-thirds of the tongue Loss of taste on the palate on the affected side Loss of salivary secretions from the submandibular and sublingual glands
Upper part of facial canal	Main trunk of the facial nerve Chorda tympani	Infections Bell's palsy	Facial paralysis Loss or impairment of taste on the anterior two-thirds of the tongue Loss of salivary secretions from the submandibular and sublingual glands
Lower part of facial canal	Main trunk of the facial nerve	Infections Bell's palsy	Facial paralysis
Parotid gland	Main trunk of the facial nerve or its individual branches (e.g., temporal, zygomatic, buccal, marginal mandibular, or cervical)	Parotid tumor Parotitis (e.g., mumps) Direct trauma Parotidectomy	Complete or partial facial paralysis dependent on the branch(es) involved

INTERESTING CLINICAL QUESTIONS

Q:

Why is there a difference in the muscles affected dependent on the site of pathology?

A:

If the whole side of the face is paralyzed, the lesion is peripheral.

If the forehead is spared on the side of the paralysis, the lesion is central (e.g., stroke).

The reasoning behind this is that the upper portion of the facial motor nucleus receives both crossed and uncrossed supranuclear fibers supplying the frontalis (forehead muscle) and orbicularis oculi (eye muscle) thus sparing them to a considerable degree. However, the portion of the facial motor nucleus innervating the mid and lower facial muscles does not have this dual cortical input.

Q:

Why may a lesion involving the facial nerve result in hyperacusis (sounds appearing louder than they actually are)?

A:

The condition of hyperacusis is caused by interruption of the proximal branch of the facial nerve that supplies the stapedius muscle. Normally, the stapedius muscle causes reduced ossicle movements resulting in a reduction in volume. If this muscle does not function as it should, sounds will appear louder to the patient on the side affected.

Tip!

As the branch of the facial nerve to stapedius arises proximally, this type of dysfunction indicates that the lesion is very close to the brainstem.

CHAPTER 8

Vestibulocochlear Nerve

THE ANATOMY—SUMMARY

The vestibulocochlear nerve is the eighth cranial nerve. It is a special sensory nerve (and deals with information related to hearing and balance (equilibrium)). It has two components:

1. Vestibular nerve: This part deals with information related to equilibration, as it is distributed to the saccule and utricle, as well as to the ampullary crests of the semicircular ducts.
2. Cochlear nerve: This part deals with hearing and is distributed to the hair cells of the spiral organ.

THE ANATOMY—IN MORE DETAIL

The vestibulocochlear nerve arises from the pontomedullary junction behind the facial nerve. Although there are two different central connections for the vestibulocochlear nerve, dependent on what component is being served, the main function is in the transmission of afferent information from the inner ear to the brain.

The vestibular ganglion is associated with the vestibular nerve and the spiral ganglion with the cochlear nerve. Both of these ganglia associated with the vestibulocochlear nerve are bipolar cells where the central portion goes to the brain, and the peripheral portion goes to the inner ear.

The vestibulocochlear nerve passes through the posterior cranial fossa to enter the petrous temporal bone at the internal auditory (acoustic) meatus (Figures 8.1–8.3). Within the internal auditory meatus, the vestibulocochlear nerve is accompanied by the two branches of the facial nerve (motor nerve and nervus intermedius, see Chapter 7) and the labyrinthine vessels. At this point, numerous accounts have been given in relation to communication between the vestibulocochlear and adjacent facial nerves (Paturet, 1951; Fisch, 1973; Shoja et al., 2014). Towards the lateral end of the internal auditory meatus, it

Clinical Anatomy of the Cranial Nerves. DOI: http://dx.doi.org/10.1016/B978-0-12-800898-0.00008-7

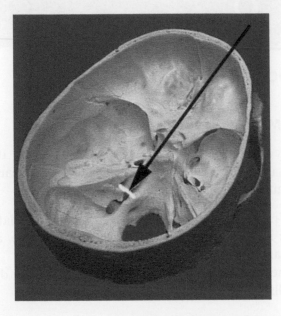

Figure 8.1 The vestibulocochlear nerve is highlighted in yellow as it passes into the internal acoustic (auditory) meatus of the petrous temporal bone, like the facial nerve (and nervus intermedius).

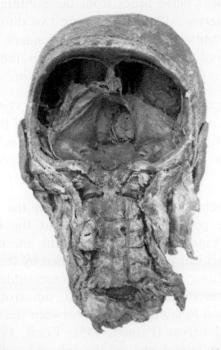

Figure 8.2 An unlabeled image to demonstrate the position of the vestibulocochlear nerve in relation to other structures.

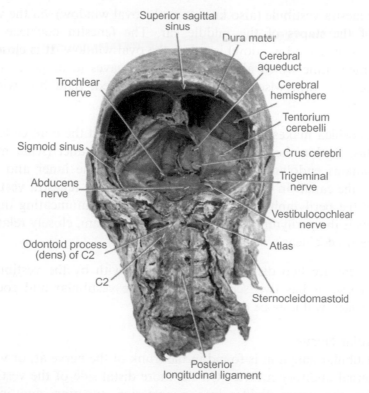

Superior sagittal sinus

Dura mater

Cerebral aqueduct

Trochlear nerve

Cerebral hemisphere

Tentorium cerebelli

Sigmoid sinus

Crus cerebri

Trigeminal nerve

Abducens nerve

Vestibulocochlear nerve

Odontoid process (dens) of C2

Atlas

C2

Sternocleidomastoid

Posterior longitudinal ligament

Figure 8.3 A labeled image to demonstrate the position of the vestibulocochlear nerve in relation to other structures.

divides into the more anterior cochlear nerve and the more posterior vestibular nerve, supplies the saccule and utricle, as well as the ampullary crests of the semicircular ducts.

Within the petrous temporal bone, the inner ear is located. The inner ear has two functions—hearing and balance (equilibration). The inner ear is comprised of the membranous labyrinth within the bony labyrinth. Sensory information for the maintenance of equilibrium comes from three systems—the eyes, proprioceptive endings throughout the entire body, and the vestibular apparatus of the inner ear. The static labyrinth, comprised of the utricle and saccule (endolymph dilations of the membranous labyrinth), detects the position of the head with respect to gravity. The kinetic labyrinth detects movement of the head via the three semicircular canals.

For hearing, the cochlear part of the inner ear contains the organ of Corti, also referred to as the spiral organ. Sound waves are transmitted

to the fenestra vestibule (also known as the oval window) via the vibra-
tions of the stapes of the middle ear. The fenestra cochleae (also
known as the round window) is below the oval window. It is closed by
a thin membrane which allows for pressure waves to be generated in
the inner ear, otherwise it would just be a fixed, rigid box with no
movement.

The cochlea makes two-and-a-half turns around the core called the
modiolus. Within the cochlea, there is a cochlear duct (scala media)
that contains endolymph and is firmly fixed to the inner and outer
walls of the canal. In close communication with the fenestra vestibule,
there is the perilymph in the scala vestibule. Communicating directly
with this is the perilymph found in the scala tympani, closely related to
the fenestra cochleae.

As there are two distinct functions dealt with by the vestibulocochlear nerve, it has two separate nerves: the vestibular and cochlear
nerves which will now be dealt with in turn.

Vestibular Nerve

The vestibular ganglion is found in the trunk of the nerve at, or within,
the internal auditory meatus. On the more distal side of the vestibular
ganglion, the nerve divides into a posterior, superior, and inferior
branches. The fibers are distributed as follows:

1. Posterior division—Passes to the ampullary crest of the posterior
   semicircular duct
2. Superior division—Passes to the macula of the utricle and the
   ampullary crests of the lateral and anterior semicircular ducts
3. Inferior division—Passes to the macula of the saccule.

Cochlear Nerve

The spiral ganglion is found at the spiral canal of the modiolus. The
fibers from here pass to the edge of the osseous spiral lamina. Some
fibers pass to the outer hair cells, while most pass to the basal and
middle coils.

From the spiral ganglion, two cell types have been identified
(Spoendlin, 1988). These are described as follows:

- Type I—Large bipolar myelinated cells with long axons projecting
  centrally and peripherally. Type I cells account for the majority of

cochlear nerve cells and are for the inner hair cells. Ten ganglion cells are connected to each sensory cell.

- Type II—Small nonmyelinated cells with a peripherally directed process. These are smaller in number (approximately 10%) and are afferent for the outer hair cells.

Therefore, the funnel-shaped external auditory canal collects sound. The purpose of this structure is to amplify the sound transmitting it to the tympanic membrane. Then, the malleus, incus, and stapes amplify this signal, and allow the sound transmission to be converted to a mechanical effect due to the vibration of the tympanic membrane. The middle ear also dampens any excessive vibration and therefore allows for impedance matching. The force at the stapes per unit area of oscillating surface is increased some 20-fold (Williams and Warwick, 1980). The basilar membrane of the inner ear then receives this information and is involved in mechanical and neural filtering and analysis of signals by the spiral organ. The inner ear and basilar membrane are also involved in stimulus transduction, and initiation of the action potentials between the cochlear nerve cells and the sensory neurons. This then allows for the conversion of sound into something electrical, and that the higher brain functions can become involved in interpreting.

The Components

The vestibulocochlear nerve comprises special somatic afferent information required for conveying information related to hearing and equilibrium.

The Ganglia

Vestibular Ganglion

The ganglion associated with the vestibular nerve is the vestibular ganglion, also known as the ganglion of Scarpa.

The centrally projecting fibers of the vestibular nerve pass to the brain above and medial to those of the cochlear nerve. They pass to the pons dividing into ascending and descending branches, mostly ending in the vestibular nuclei.

Four vestibular nuclei are described as the lateral, superior, medial, and inferior nuclei. From these nuclei, there are essentially three pathways—that related to the cerebellum, that connected with the spinal cord, and that connected with the brainstem.

Vestibulocerebellum

The superior, medial, and inferior vestibular nuclei constitute part of the vestibulocerebellum. Afferent and efferent fibers exist between the cerebellum and the vestibular nerve and function in maintaining communication between these two sites. It helps regulate balance and related eye movements and is found in the flocculonodular lobe.

Vestibulospinal Tract

The lateral vestibular nucleus, also referred to as Deiter's nucleus, constitutes the fibers which pass towards the spinal cord, forming the vestibulospinal tract. The vestibulospinal tract then passes to the ventral horn, primarily to lamina VIII but to a lesser degree lamina VII. This is most abundant at the cervical and lumbosacral territories. The vestibulospinal tract is for maintenance of balance by regulation of the muscle tone related to posture.

Other fibers that travel within the spinal cord extend from the medial vestibular nucleus projecting in the medial longitudinal fasciculi of both sides of the body. These fibers are concerned in interacting with the cervical motor neurons, and therefore the neck musculature, to maintain balance and the fixation of the eyes.

Brainstem Connections

The vestibular nuclei are connected with the oculomotor, trochlear, and abducent nerves via the ascending part of the medial longitudinal fasciculus. This pathway is both crossed and uncrossed and is concerned with movement of the head in a coordinated fashion to maintain visual fixation, but is also concerned with conjugate gaze.

Spiral Ganglion

At the periphery of the modiolus, and in a spiral arrangement (hence its name), is the spiral ganglion. Peripheral processes from here extend to the organ of Corti. The central processes from the spiral ganglion then synapse several times, and in different locations, on their way to the primary auditory area of the cerebral cortex in the superior temporal gyrus. The first port of call for fibers from the spiral ganglion is either the *ventral or dorsal cochlear nuclei*. Some nerve fibers that have their cell bodies in the spiral ganglion will synapse in the *ventral cochlear nucleus*. They can then either pass via one of two routes. The first route from the ventral cochlear nucleus after synapsing there is to then pass to the *ipsilateral superior olivatory nucleus*. Here, they will synapse and pass to the *ipsilateral inferior colliculus* to synapse again. From here, and without crossing, they will pass to synapse in the *medial geniculate body* and then onwards to terminate in the *ipsilateral superior temporal gyrus*. Other fibers which synapse in the *ventral cochlear nucleus* pass to the *contralateral superior olivatory nucleus* to synapse, and then via the *lateral lemniscus*, have the same route as previously described. The only difference is that the pathway is on the opposite side to where the spiral ganglion is located.

Other fibers can pass from the spiral ganglion to synapse in the *dorsal cochlear nucleus*. They then pass to the *opposite* superior olivatory nucleus to synapse in the inferior colliculus and medial geniculate body, terminating on the opposite sides of the primary auditory cortex in the superior temporal gyrus.

THE CLINICAL APPLICATION

Testing at the Bedside
Basic Testing
Assess hearing by instructing the patient to close their eyes and to say "left" or "right" when a sound is heard in the respective ear. Vigorously rub your fingers together very near to, yet not touching, each ear and wait for the patient to respond. After this test, ask the patient if the sound was the same in both ears or louder in a specific ear. If there is lateralization or hearing abnormalities perform the Rinne and Weber tests using the 512 Hz tuning fork.

Weber Test
The Weber test is a test for lateralization. Tap the tuning fork strongly on your palm and then press the butt of the instrument on the top of the patient's head in the midline and ask the patient where they hear the sound. Normally, the sound is heard in the center of the head or equally in both ears. If there is a conductive hearing loss present, the vibration will be louder on the side with the conductive hearing loss. If the patient doesn't hear the vibration at all, attempt again, but press the butt harder on the patient's head.

Rinne Test

The Rinne test compares air conduction to bone conduction. Tap the tuning fork firmly on your palm and place the butt on the mastoid eminence firmly. Tell the patient to say "now" when they can no longer hear the vibration. When the patient says "now," remove the butt from the mastoid process and place the U of the tuning fork near the ear without touching it.

Tell the patient to say "now" when they can no longer hear anything. Normally, one will have greater air conduction than bone conduction and therefore hear the vibration longer with the fork in the air. If the bone conduction is the same or greater than the air conduction, there is a conductive hearing impairment on that side. If there is a sensorineural hearing loss, then the vibration is heard substantially longer than usual in the air.

Make certain that you perform both the Weber and Rinne tests on both ears. It would also be prudent to perform an otoscopic examination of both eardrums to rule out a severe otitis media, perforation of the tympanic membrane or even occlusion of the external auditory meatus, which all may confuse the results of these tests. If hearing loss is noted, an audiogram is indicated to provide a baseline of hearing for future reference.

Otoscopy

This will allow the external auditory meatus and the middle ear to be assessed by examination of the tympanic membrane.

Advanced Testing
Audiometry (Hearing) Tests

Automated Otoacoustic Emission (AOAE) Test

The AOAE test works on the basis that the normal cochlea, when stimulated by sound, produces an echo. Therefore, this test involves an earpiece with a speaker and microphone connected. Clicking sounds are played through the speaker, and if the cochlea is functioning, the microphone will detect the echo.

This is a simple and effective method to quickly assess the hearing of a patient, including that of the very young patient. It allows for an immediate result to be given and from this it can then be classified as "pass" or "refer." Refer does not necessarily mean that there is a gross abnormality. It may simply be that there was too much background noise in the room interfering with the test, or in the very young patient, fluid from childbirth may still have been present in the external auditory meatus, or the child may have just been too restless. More formal testing can therefore help to give a clearer indication of the patient's hearing status.

Automated Auditory Brainstem Response (AABR) Test

Sensors are placed on the patient's head and neck. It works by sounds being played through headphones which the patient wears for the duration of the test. The sensors will detect these sounds being transmitted. A small percentage of patients (generally babies) can be referred on for a more in-depth analysis of their hearing responses.

Pure Tone Audiometry (PTA) Test

This test is one of the key hearing tests, but only suitable for adults and children old enough to be able to communicate effectively and cooperate with the test. This test tests both air and bone conduction. In a sound-proof room, the patient has headphones placed on, as well as a headband to assess bone conduction. A variety of frequencies and volumes are then played and the patient then has to press a button on a machine to indicate when they hear the noise. The headband can assess the bone conduction.

Bone Conduction Tests

This test is a more advanced form of using tuning forks, and combined with a PTA test, allows differentiation of external, middle, or inner ear pathology.

A variety of other formal tests are also available in assessing hearing including tympanometry, electrophysiological, and acoustic reflex testing. For specialist advice, it is wise to consult your local audiologist and/or otorhinolaryngologist.

Vestibular Tests

Rotation Testing

This test assesses eye movements when the head is rotating at a variety of speeds. Two types exist—auto head rotation and rotary chair. With the former, the patient moves their head back and forward observing a fixed target. With the rotary chair, a computer will control the movement of the chair and the eye movements are recorded.

Electronystagmography (ENG)
This test assesses nystagmus with electrodes placed on the skin around the eye. The other method for assessing this is to use videonystagmography where an infrared camera in glasses monitors eye movements. Overall, the test observes the eye either in a fixed position in different directions or having the eyes follow a moving target.

Computerized Dynamic Posturography (CDP)
This test assesses postural stability by standing on a platform with a visual target to observe. The visual target and/or the platform will move and the three inputs for posture are therefore assessed, i.e., visual, vestibular, and sensory information from the joints and muscles.

Vestibular Evoked Myogenic Potential (VEMP)
By placing headphones over the ears and electrodes on the skin overlying the neck musculature, a sound impulse is played. The purpose of this study is to assess the response of vestibular stimuli and assesses the inferior vestibular nerve and the saccule.

A variety of hearing tests can be undertaken as described in the "Audiometry (Hearing) Tests" section. As directed by the patient's history and examination, other tests could be an MRI, which will aid in the diagnosis of any soft tissue problem, and/or CT which will help in the diagnosis perhaps of temporal bone pathology.

PATHOLOGIES

A number of pathologies can affect the vestibulocochlear nerve as detailed below.

Deafness
Two different types of deafness exist—conductive and sensorineural.

Conductive Deafness
With conductive deafness, there is impaired transmission of the sound waves along the external auditory meatus to the middle ear (stapes). A variety of causes can result in conductive deafness like ear wax, discharge from otitis externa, otitis media (supportive and serous), damage

to the tympanic membrane (perhaps from rupture or scarring), foreign body, or congenital abnormality.

Sensorineural Deafness

Sensorineural deafness is caused by pathology central to the oval window in the cochlea (therefore sensory), cochlear part of the vestibulocochlear nerve (therefore neural), or perhaps within central pathways. The most common cause of this is exposure to loud noises, especially over prolonged periods of time. Other notable causes of sensorineural hearing loss are infections/inflammation, trauma (base of skull fracture), ischemia, and ototoxic drugs, e.g., gentamycin.

Vertigo

Vertigo is a feeling of disorientation and can include feeling dizzy, or that the environment is moving or spinning. Nausea and vomiting, nystagmus, and hearing loss can accompany it. The causes of vertigo are described as either being central or peripheral in origin. Central causes can include neurological (multiple sclerosis), vascular (vertebrobasilar ischemia), infectious (syphilis, herpes), trauma to the head or tumor (acoustic neuroma). Peripheral causes can include vestibular neuronitis, cholesteatoma, labyrinthitis, benign postural vertigo, or Ménière's disease.

Tinnitus

Tinnitus is where the patient will experience ringing or buzzing in the ears. The exact mechanism is uncertain; however, related causes can be trauma, drugs (e.g., aspirin, loop diuretics, and gentamycin), infection (suppurative otitis media and viral), and psychological issues.

Acoustic Neuroma

An acoustic neuroma is also called a vestibular schwannoma. This tumor arises from the Schwann cells. It is a slow-growing tumor, growing at approximately 1−2 mm per year. As it is a slow-growing tumor, it takes some time before signs and symptoms present. When they do present, hearing loss will be the most obvious symptom. It is a sensorineural deafness and can also present as vertigo, nausea, altered balance, and the majority of patients will have tinnitus. The cause of an acoustic neuroma is generally unknown but can be related to von Recklinghausen neurofibromatosis.

INTERESTING CLINICAL QUESTIONS

Q:
What is the vestibulo-ocular reflex?
A:
This is a reflex eye movement which results in stabilizing the image which falls onto the retina by moving the eye in the opposite direction to which the head moves.

Q:
What would the vestibule-ocular reflex test show in a patient with brainstem death?
A:
To assess this, the eyes would be observed while gently turning the patient's head. In the comatose patient with a normal vestibule-ocular reflex, the eyes would move to the opposite side, i.e., rotating the head to the left would result in the eyes moving to the right. This is normally seen as though the eyes are still fixed on the examiner. If there was brainstem death, the eyes would stay fixed mid-orbit.

Q:
What is caloric testing?
A:
Caloric testing or the caloric reflex test is a test of the vestibule-ocular reflex (previously described) where warm or cold water is gently infused into the external auditory meatus. Warm or cold water is used to ensure stimulation outside the normal body temperature range. They have opposite results which has to be remembered when

testing! If warm water is injected, both eyes rotate to the opposite side with the nystagmus (horizontal) to the ipsilateral ear of the injection. However, if cold water is injected, the eyes initially turn to the side of injection with the horizontal nystagmus to the opposite side. This is because the warm water will cause an increase in the vestibular nerve firing while the opposite is true of the cold water. These results will only be present in someone with an intact brainstem.

REFERENCES

Fisch, U.P., 1973. Excision of the Scarpa's ganglion. Arch. Otolaryngol. 97, 147—149.

Paturet, G., 1951. Traite D'anatomie Humaine, vol. 1. Masson et Cie, Paris.

Shoja, M.M., Oyesiku, N.M., Griessenauer, C.J., Radcliff, V., Loukas, M., Chern, J.J., et al., 2014. Anastomoses between lower cranial and upper cervical nerves. Clin. Anat. 27, 118—130.

Spoendlin, H., 1988. In: Alberti, P.W., Ruben, R.J. (Eds.), Biology of the Vestibulocochlear Nerve, first ed. Churchill Livingstone, New York, NY, pp. 117—150.

Williams, P.L., Warwick, R., 1980. Gray's Anatomy, thirty-sixth ed. Churchill Livingstone, New York, NY.

CHAPTER 9

Glossopharyngeal Nerve

THE ANATOMY—SUMMARY

The glossopharyngeal nerve is the ninth cranial nerve. It is rather complex having four related nuclei conveying information related to sensation, muscular activity, and autonomic function. It contains special visceral efferent (visceral motor), general visceral efferent (parasympathetic), general visceral afferent (visceral sensory), and general somatic afferent (general sensory) information from a variety of structures.

THE ANATOMY—IN MORE DETAIL

The glossopharyngeal nerve contains sensory fibers from the pharynx, tongue (posterior one-third), and the tonsils. It also contains secretomotor fibers destined for the parotid gland as well as motor fibers for the stylopharyngeus. Finally, it also contains taste fibers, from the posterior one-third of the tongue. Therefore, the glossopharyngeal nerve is a rather complex and important nerve supplying a variety of structures.

The glossopharyngeal nerve arises as three or four rootlets at the level of the medulla oblongata. It passes out from between the inferior cerebellar peduncle and the olive, superior to the rootlets of the vagus nerve. It then sits on the jugular tubercle of the occipital bone. It then runs to the jugular foramen, passing through the middle part of it (Figure 9.1). At the point of entry to the jugular foramen, two ganglia are found—an inferior ganglion and a superior ganglion. Both of these ganglia contain the cell bodies of the afferent fibers contained within the glossopharyngeal nerve. On passing through the jugular foramen, the glossopharyngeal nerve then passes between the internal carotid artery and the internal jugular vein, descending in front of the artery. It then passes deep to the styloid process and related muscles attaching on to this bony prominence. It then winds round the stylopharyngeus, passing deep to the hyoglossus and going between the superior and middle pharyngeal constrictors (Figures 9.2 and 9.3).

Clinical Anatomy of the Cranial Nerves. DOI: http://dx.doi.org/10.1016/B978-0-12-800898-0.00009-9

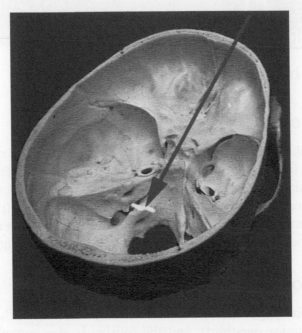

Figure 9.1 The glossopharyngeal nerve will leave the skull through the jugular foramen as highlighted by the outline of the nerve's passage under the petrous temporal bone.

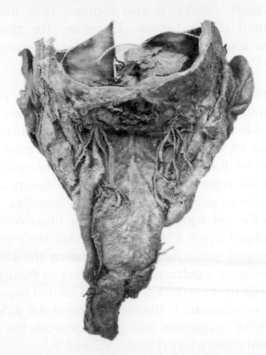

Figure 9.2 An unlabeled image to demonstrate the position of the glossopharyngeal nerve in relation to other structures.

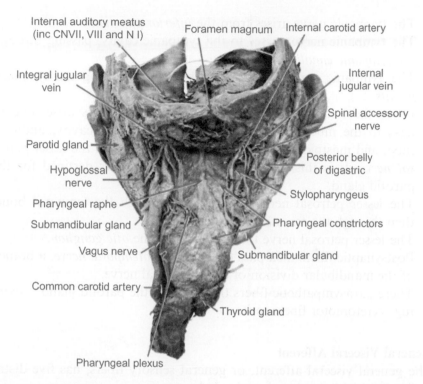

Figure 9.3 A labeled image to demonstrate the position of the glossopharyngeal nerve in relation to second structures.

The Components
Special Visceral Efferent

The glossopharyngeal nerve contains branchial motor fibers to the sty-lopharyngeus muscle, which is derived from the third pharyngeal arch. This branch is given off to the muscle as it passes across it as it descends down from the jugular foramen. The stylopharyngeus muscle is responsible for raising the larynx and pharynx and functions during swallowing.

General Visceral Efferent (Autonomic—Parasympathetic)

The general visceral efferent, or parasympathetic fibers of the glosso-pharyngeal nerve, initially arises from the tympanic nerve. The following sequence then occurs for the tympanic nerve to supply the parotid gland, as detailed below.

1. The tympanic nerve arises from the *inferior ganglion*.
2. The tympanic nerve passes to the tympanic cavity, passing through the *tympanic canaliculus*.
3. The tympanic nerve divides into branches forming the *tympanic plexus*.
4. From the tympanic plexus, two branching patterns arise: one to pass to the mucous membranes of the tympanic cavity, auditory tube, and mastoid air cells. The other branch gives the *lesser petrosal nerve*. It is this nerve that contains the fibers destined for the parotid gland.
5. The lesser petrosal nerve then passes through the temporal bone, then through the *foramen ovale*.
6. The lesser petrosal nerve then synapses in the *otic ganglion*.
7. Postsynaptic fibers then pass to the *auriculotemporal nerve*, a branch of the mandibular division of the trigeminal nerve.
8. These parasympathetic fibers then pass to the parotid gland providing secretomotor fibers to it.

General Visceral Afferent
The general visceral afferent, or general sensory fibers, has five distributions as highlighted in the table below.

Branch	Distribution
Carotid sinus branch	This branch passes to the anterolateral aspect of the internal carotid artery It passes to the baroreceptors and chemoreceptors in the carotid sinus and the carotid body
Tympanic nerve	The sensory fiber component conveys information from the middle ear
Lingual branch	Conveys sensory information from the posterior one-third of the tongue as well as the vallate papillae
Tonsillar branch	Conveys information from the mucous membranes of the palatine tonsil and soft palate
Pharyngeal branch	This branch conveys sensory information from the oropharynx

Special Visceral Afferent
The special visceral afferent fibers contained in the glossopharyngeal nerve are concerned with taste sensation from the posterior one-third of the tongue. The sensory information passes from the posterior one-third of the tongue to the pharyngeal branches of the glossopharyngeal nerve to pass to the inferior ganglion. From here, the fibers

pass to the nucleus solitarius in the medulla where they synapse. From here, the fibers pass bilaterally to the thalamus via the ventral posteromedial nuclei and then onward to the gustatory cortex within the parietal lobe.

General Somatic Afferent
The general somatic afferent, or general sensory fibers, conveys general sensory information from the skin of the external ear, inside of the tympanic membrane, the upper portion of the pharynx as well as general sensation from the posterior one-third of the tongue.

The fibers from the skin of the external ear initially travel with the vagus nerve (auricular branch (Arnold's nerve)). From the inner aspect of the tympanic membrane, the fibers for general sensation travel in the tympanic nerve. The fibers for general sensation from the upper part of the pharynx and posterior one-third of the tongue pass via the pharyngeal branch of the glossopharyngeal nerve.

From these branches, they then pass centrally to the medulla entering the spinal nucleus of the trigeminal nerve, projecting contralaterally to the ventral posteromedial nucleus of the thalamus. From there, the fibers from the external ear, tympanic membrane, pharynx, and tongue then terminate in the sensory cortex for interpretation and processing of the information conveyed.

The Ganglia
Superior ganglion—This ganglion is very small and is sometimes viewed as a broken off part of the inferior ganglion. It is found within the groove of where the glossopharyngeal nerve passes in the jugular foramen. It contains the visceral sensory fibers from the pharynx, parotid gland, carotid body, and sinus as well as the middle ear.

Inferior ganglion—The inferior ganglion conveys information related to special and general sensation from the mucous membrane of the posterior one-third of the tongue. Its peripheral fibers also come from the oropharynx and soft palate conveying general sensory fibers. The inferior ganglion is the bigger of the two ganglia related to the glossopharyngeal nerve and is found on the lower border of the petrous temporal bone in a notch. The inferior ganglion also communicates with the facial and vagus nerves, as well as with the sympathetic trunk.

The Important Branches
The following table highlights the important branches of the glossopharyngeal nerve and what the functions of each of these branches are
responsible for:

Branch	Function
Muscular	Motor to stylopharyngeus
Tympanic	Parasympathetic fibers to parotid gland and general sensation to middle ear
Lingual	General sensation from the tongue (posterior one-third) and vallate papillae. May contain taste fibers from the posterior one-third of the tongue
Pharyngeal	Special sensory fibers to the posterior one-third of the tongue. General sensation to the oropharynx
Tonsillar	General sensation from the soft palate and palatine tonsil
Carotid sinus	Visceral sensory fibers from the carotid body and sinus

THE CLINICAL APPLICATION

Testing at the Bedside

1. The first thing to say is that examination of the glossopharyngeal
   nerve is difficult. Assessing it on its own is not possible, and an isolated lesion of this nerve is almost unknown (Walker et al., 1990).
   When assessing the glossopharyngeal nerve, the first thing to do is
   simply *listening to the patient talking.* Any abnormality of the voice,
   for example, hoarse, whispering, or a nasal voice, may give a clue
   as to an abnormality. Also, ask the patient if they have any difficulty in swallowing. The result of a glossopharyngeal nerve (and
   related cranial nerves, e.g., vagus and accessory nerves, due to their
   close proximity to each other) may be dysphagia (difficulty swallowing), aspiration pneumonia, or dysarthria (difficulty in the
   motor control of speech).
2. To assess the function of the glossopharyngeal nerve (and the vagus
   nerve), ask the patient to say "ahhhh" (without protruding their
   tongue) for as long as they can. Normally, the palate should rise
   equally in the midline. The palate (uvula) will move *toward* the side
   of the lesion if there is a problem with the glossopharyngeal (and
   perhaps vagus) nerve.
3. Damage to the glossopharyngeal (and vagus) nerve for example
   from a stroke may result in loss of the gag reflex. ALWAYS tell the

patient what you will do before assessing the gag reflex, as it is not a pleasant examination, and may not always be necessary.
4. A swab can be used to gently touch the palatal arch on the left- then right-hand sides. Try to assess the normal side first if you suspect a pathology.

Tip!

If it is difficult to view the palate (and uvula) when examining the glosso-pharyngeal (and vagus) nerve, a tongue depressor allows for easier visual-ization. Again, always tell the patient what you will be doing before doing it, as some patients do not like the taste/texture of a tongue depressor.

Advanced Testing

It may be that taste from the posterior one-third of the tongue will need to be tested, though is not commonly performed. This should be done in the same way that the facial nerve is tested for taste. ALWAYS tell the patient what you will be doing before examining them. This allows the patient to be fully informed and consent to the procedure but also will build up trust between you and the patient.

A small tongue blade can be used to place a small amount of salt or sugar onto each part of the tongue needed testing. It should be placed on the lateral aspect of the tongue. Remember that the glosso-pharyngeal nerve only supplies the posterior one-third of the tongue for taste sensation.

PATHOLOGIES

Isolated glossopharyngeal nerve palsy is extremely rare. Indeed, unilateral lesions of the glossopharyngeal nerve tend not to cause major deficits, as there is bilateral corticobulbar input. A bilateral lesion will result in a pseudobulbar palsy. This is described in more detail in Chapter 10. A variety of causes of glossopharyngeal nerve palsy can be from intracranial tumors (cerebellopontine angle) or neck tumors.

INTERESTING CLINICAL QUESTIONS

Q:
What is glossopharyngeal neuralgia, and how would it present?
A:
This is a rare condition, with an unknown cause, although may be related to compression of the glossopharyngeal nerve by a nearby blood vessel, base of skull tumor around the course of the nerve, or a tumor within the throat compressing on its extracranial part. It typically presents with pain at the back of the throat, tongue, and ear. These episodes are triggered by a variety of activities involving using the mouth (e.g., eating/chewing, swallowing, laughing, and speech). The episodes tend to get worse as time progresses. If it involves compression of the vagus nerve too, it can present with pain and bradycardia and in extreme cases asystole. Syncope may also be encountered with vagus nerve involvement. If there is a treatable cause, that should be managed first, as well as providing analgesia and perhaps, as with trigeminal neuralgia, an antiepileptic agent like carbamazepine.

Q:
What is jugular foramen syndrome?
A:
Jugular foramen syndrome, also known as Vernet's syndrome, is a condition which affects the glossopharyngeal, vagus, and accessory nerves as they enter the jugular foramen. Compression of these nerves, and perhaps also the hypoglossal nerve, is caused by a variety of neoplasms, trauma, and infections. It leads to a wide variety

of signs and symptoms. Typically, this will involve loss of taste to the posterior one-third of the tongue, dysphagia, vocal paralysis and anesthesia of the larynx and pharynx, and weakness or paralysis of the trapezius and sternocleidomastoid muscles (Greenberg, 2010). Treatment is aimed at an identifiable cause.

REFERENCES

Greenberg, M.S., 2010. Handbook of Neurosurgery. Thieme, New York, NY, ISBN-10: 1604063262.

Walker, H.K., Hall, W.D., Hurst, J.W., 1990. Clinical methods, The History, Physical, and Laboratory Examinations, third ed. Butterworths, Boston, MA, ISBN-10: 0-409-90077-X. Found at: <http://www.ncbi.nlm.nih.gov/books/NBK201/> (accessed 09.01.14.).

CHAPTER 10

Vagus Nerve

THE ANATOMY—SUMMARY

The vagus nerve is the tenth cranial nerve. Like the glossopharyngeal nerve, again, it is a rather complex nerve having four nuclei and five different types of fibers in it. These convey information related to sensory, muscular activity, and autonomic functions. Its name comes from the Latin word *vagary*, meaning wandering. It has the longest course of the cranial nerves and is extensively distributed, especially below the level of the head. It contains the following types of fibers:

a. Branchial motor
   Supplying muscles of the pharynx and larynx.
b. Visceral sensory
   This component of the vagus nerve is responsible for transmitting information from a wide variety of anatomical sites including the heart and lungs, pharynx and larynx, and upper part of the gastrointestinal tract.
c. Visceral motor
   The visceral motor component carries parasympathetic fibers from the smooth muscle of the upper respiratory tract, heart, and gastrointestinal tract.
d. Special sensory
   The special sensation conveyed by the vagus nerve is for taste from the palate and epiglottis.
e. General sensory
   The general sensory component of the vagus nerve is concerned with information from parts of the ear and the dura within the posterior cranial fossa.

THE ANATOMY—IN MORE DETAIL

The vagus nerve arises from the medulla as several rootlets. It passes toward the jugular foramen (Figure 10.1) found between the glossopharyngeal and spinal accessory nerves. Like the glossopharyngeal nerve,

Clinical Anatomy of the Cranial Nerves. DOI: http://dx.doi.org/10.1016/B978-0-12-800898-0.00010-5

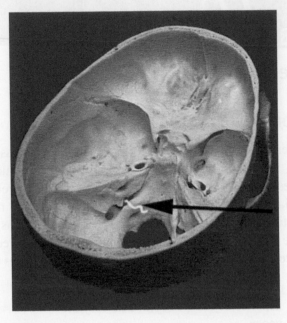

Figure 10.1 The passageway of the vagus nerve is shown as the yellow structure, highlighted with the arrow. It passes into the jugular foramen to exit the skull.

two ganglia related to the vagus nerve are found here—the superior and inferior ganglia. The vagus nerve then descends in the carotid sheath sandwiched between the internal jugular vein and the internal and external carotid arteries. As it further descends, it is related to the internal jugular vein and the common carotid artery. Then, the right and left vagus nerves have very different anatomical pathways.

On the right side, the vagus nerve passes anterior to the right subclavian artery and posterior to the superior vena cava. At the point where it is closely related to the subclavian artery, it gives off its recurrent laryngeal branch. This branch passes under the artery then posterior to it. It then ascends between the trachea and esophagus, both of which it supplies at that point. The right recurrent laryngeal nerve then passes closely related to the inferior thyroid artery. It enters the larynx behind the cricothyroid joint and deep to the inferior constrictor. The recurrent laryngeal nerve conveys sensory information from below the level of the vocal folds, and all of the muscles of the larynx on that side, except cricothyroid.

The left vagus nerve descends toward the thorax passing between the common carotid and subclavian arteries, passing posterior to the

Figure 10.2 An unlabeled image to demonstrate the position of the vagus nerve in relation to other structures.

brachiocephalic vein. It gives off branches here to the esophagus, lungs, and heart. It then passes to the left side of the arch of the aorta. From here, the recurrent laryngeal nerve is given off which descends underneath the arch of the aorta to ascend in the groove between the esophagus and trachea. As it does so, it gives off branches to the aorta, heart, esophagus, and trachea (Figures 10.2–10.5).

The Components
The vagus nerve has five main components contained within the nerve. These are branchial motor, visceral sensory, visceral motor, and special and general sensory fibers.

Branchial Motor
The branchial motor component supplies all of the muscles of the pharynx, larynx, and the soft palate. There are several in number and called the pharyngeal branches of the vagus nerve. The only muscles that are not supplied in this territory by the vagus nerve are the stylopharyngeus (supplied by the glossopharyngeal nerve) and the tensor veli palatini (supplied by the medial pterygoid nerve from

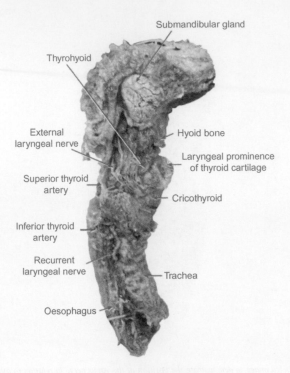

Figure 10.3 A labeled image to demonstrate the position of the vagus nerve in relation to other structures.

Figure 10.4 An unlabeled image to demonstrate the position of the vagus nerve in relation to other structures.

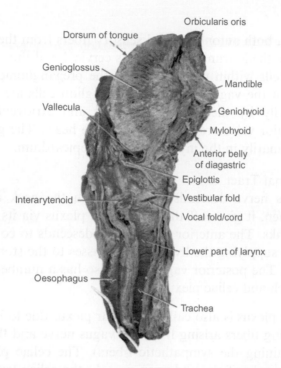

Orbicularis oris

Dorsum of tongue

Genioglossus

Mandible

Vallecula

Geniohyoid

Mylohyoid

Anterior belly
of diagastric

Epiglottis

Interarytenoid

Vestibular fold

Vocal fold/cord

Lower part of larynx

Oesophagus

Trachea

Figure 10.5 A labeled image to demonstrate the position of the vagus nerve in relation to other structures.

the mandibular division of the trigeminal nerve). One point to note is that the recurrent laryngeal nerve, as well as containing sensory fibers in it for below the level of the vocal folds, supplies all the laryngeal muscles apart from the cricothyroid muscle. A separate branch of the superior laryngeal nerve called the external laryngeal nerve supplies this muscle. This will be discussed in "The Important Branches."

Visceral Motor

The visceral motor component of the vagus nerve contains the parasympathetic part of this nerve. It supplies a wide variety of structures from the head and neck through to the thorax and abdomen. The vagus nerve supplies parasympathetic innervation to the mucous membranes of the pharynx and larynx, trachea, bronchia, and the gastrointestinal tract. Simply put, stimulation of the vagus nerve would slow heart rate, cause bronchoconstriction, and result in improved gastrointestinal contractility.

Heart

For the heart, both autonomic and sensory fibers from the vagus nerve and the sympathetic trunk supply it. Its cervical and thoracic branches to ganglion cells within the heart carry the preganglionic parasympathetic fibers of the vagus nerve. Many ganglion cells are found in the atria, especially by the nodes (sinuatrial and atrioventricular) and where the major veins are emptying into the heart. The ganglion cells are found primarily in the myocardium and epicardium.

Gastrointestinal Tract

As the vagus nerve descends down through the neck and into the upper abdomen, it gives off the esophageal plexus via its anterior and posterior trunks. The anterior vagal trunk descends to course with the hepatic atery supplying the liver. It also passes to the stomach and the celiac plexus. The posterior vagal trunk also has a number of branches to the stomach and celiac plexus.

The celiac plexus is also called the solar plexus due to its huge number of radiating fibers arising from the vagus nerve and the splanchnic nerves (containing the sympathetic fibers). The celiac plexus follows the arterial supply of the abdomen namely the celiac trunk (or celiac axis), superior mesenteric, and inferior mesenteric arteries.

The vagal fibers that constitute the celiac and superior mesenteric plexuses supply parasympathetic innervation to the liver, pancreas, and the small and large intestines, as far as the last one-third of the transverse colon. The pelvic splanchnic nerves, from the second to fourth sacral segments, supply the last one-third of the transverse, descending and sigmoid colon, as well as the rectum and anal canal.

The Ganglia

Like the glossopharyngeal nerve, the vagus nerve also has two ganglia associated with it—the superior and inferior ganglia.

Superior ganglion—This ganglion is found at the level of the jugular foramen. It has motor, parasympathetic, and sensory fibers all passing through it. However, it also contains the cell bodies of general sensory fibers, specifically from the external ear.

Inferior ganglion—Motor and parasympathetic fibers pass through the inferior ganglion (also called the nodose (hard/lumpy) ganglion). However, it is primarily concerned with the visceral afferent fibers for information coming from the abdominal and thoracic viscera as well as the pharynx.

The Important Branches
These branches are listed as they arise from the superior to the inferior aspect of the vagus nerve as it passes from its point of origin (medulla) down through the neck and thorax terminating in the abdominal structures.

1. Meningeal branch
   This branch arises at the superior ganglion but contains the first and second cervical spinal nerves supplying the dura in the posterior cranial fossa.
2. Auricular branch
   This branch also arises from the superior ganglion and is joined by a branch from the glossopharyngeal nerve. It also may have a communication with a branch of the facial nerve and supplies the auricle (cranial surface) and the tympanic membrane as well as the floor of the external auditory meatus.
3. Pharyngeal branches
   These branches, of which there can be several, help form the pharyngeal plexus (which also contains the glossopharyngeal nerve and the sympathetic branches) on the constrictor muscles. They supply all the muscles of the soft palate (apart from tensor veli palatini which is supplied by the medial pterygoid nerve (from the mandibular division of the trigeminal nerve)) and pharynx (apart from stylopharyngeus which is supplied by the glossopharyngeal nerve). Most fibers are derived from the internal branch of the accessory nerve.
4. Superior laryngeal nerve
   This branch descends closely related to the pharynx and has two branches—an internal and external components. The internal laryngeal nerve supplies sensation to the mucosa from the epiglottis to just *above* the level of the vocal folds. (The recurrent laryngeal nerve supplies sensation from the rest of the larynx *below* the level of the vocal folds.) It pierces the thyrohyoid membrane above the superior laryngeal artery. The other branch of the superior laryngeal nerve, the external laryngeal nerve, passes under sternothyroid deep to the superior thyroid artery. It supplies the cricothyroid and the inferior constrictor muscles.

5. Recurrent laryngeal nerve

   The right recurrent laryngeal nerve arises from in front of the sub-clavian artery. It then ascends alongside the trachea posterior to the common carotid artery. At the inferior pole of the thyroid gland, the recurrent laryngeal nerve is closely related to the inferior thyroid artery. Like the left recurrent laryngeal nerve, both of these branches have highly variable relations to the inferior thyroid artery, and the surgeon must be cautious of this when operating in and around here (e.g., thyroidectomy) (Yalcxin, 2006).

   On the left-hand side, the recurrent laryngeal nerve arises at the arch of the aorta and then passes underneath it, closely related to the ligamentum arteriosum. It then ascends in the tracheoesopha-geal groove.

   Both recurrent laryngeal nerves pass deep to the inferior con-strictor muscle and enter the larynx at the junction between the inferior cornu of the thyroid and the cricoid cartilage. They carry several types of fibers in them—motor to all the muscles of the lar-ynx (apart from cricothyroid (supplied by the external laryngeal nerve)), sensory fibers (to below the level of the vocal folds), and stretch receptors from the larynx. As it passes close to the subcla-vian artery or aorta, it gives off cardiac branches.

6. Carotid branches

   These branches are variable in number and can arise from either the glossopharyngeal or superior laryngeal nerve, or the inferior ganglion.

7. Cardiac branches

   These branches exist as either two or three separate branches arising at the superior or inferior cervical levels. Both of these branches pass to the cardiac plexus, merging with the sympathetic branches to the heart.

8. Esophageal branches

   These branches form the esophageal plexus providing innervation to the esophagus and the posterior aspect of the pericardium.

9. Pulmonary branches

   The pulmonary branches exist as anterior and posterior branches, with the posterior fibers being more numerous. They ramify with the sympathetic branches supplying the bronchi and related pulmonary tissue.

10. Gastric branches

The left vagus nerve supplies primarily the anterosuperior aspect of the stomach. The right vagus nerve supplies the more postero inferior region of the stomach.

11. Celiac branches

These branches go on to form the celiac plexus, contributing also to the hepatic plexus supplying the liver. These fibers amalgamate with the fibers of the sympathetic trunk.

12. Renal branches

The renal branches from the vagus nerve help to contribute to the renal plexus, which includes the splanchnic nerves. These branches go on to supply the blood vessels, glomeruli, and tubules.

THE CLINICAL APPLICATION

Testing at the Bedside

Testing of the vagus nerve is done in exactly the same way that the glossopharyngeal nerve is tested.

1. When assessing the vagus nerve, as with the glossopharyngeal nerve, the first thing to do is simply *listening to the patient talking*. Any abnormality of the voice, for example, hoarse, whispering, or a nasal voice, may give a clue as to an abnormality. Also, ask the patient if they have any difficulty in swallowing.

2. To assess the function of the vagus nerve (and the glossopharyngeal nerve) ask the patient to say "ahhhh" (without protruding their tongue) for as long as they can. Normally, the palate should rise equally in the midline. The palate (uvula) will move *toward* the side of the lesion if there is a problem with the vagus (and glossopharyngeal) nerve.

3. The gag reflex can also be assessed if relevant. You MUST tell the patient what you will do before doing this test, as it is unpleasant.

4. Using a swab, GENTLY touch each palatal arch in turn, waiting each time for the patient to gag.

●●●
_____

Tip!

Vagus nerve pathology could present with the following, affecting one or all of its branches:

a. Palatal paralysis (absent gag reflex)
b. Pharyngeal/laryngeal paralysis

c. Abnormalities with the autonomic innervation of the organs it supplies (i.e., heart, stomach (gastric acid secretion/emptying), gut motility)

Glossopharyngeal nerve pathology on the other hand will affect the following:

a. Dysphagia
b. Impaired taste and sensation on the posterior one-third of the tongue
c. Absent gag reflex
d. Abnormal secretions of the parotid gland, though difficult to assess from the patient accurately

Advanced Testing

To further assess the vagus nerve, one area that can be easily assessed more formally is by using laryngoscopy. This allows for full visualization of the vocal cords and for biopsies to be taken. If swallowing is a problem for the patient, it may be relevant, dependent on the history and examination, to undertake videofluoroscopy swallow test. Cardiovascular and gastrointestinal assessment may be considered if suggested in the clinical history and presentation.

PATHOLOGIES

An isolated nerve lesion of the vagus nerve is rare. However, if the vagus nerve is affected by pathology, whether it is neurological or trauma related, it could have widespread consequences due to the wide variety of structures it supplies.

1. Pseudobulbar palsy
   Pseudobulbar palsy is caused by a wide variety of conditions (neurological and vascular) but typically results from bilateral degeneration involving cranial nerve nuclei and the corticobulbar tract (pathway connecting the brainstem with the cerebral cortex). It results in the patient having difficulty swallowing (dysphagia) and difficulty in the motor aspect of speech production.
2. Bilateral vagus nerve nucleus pathologies
   A bilateral pathology affecting the vagus nerve will result in paralysis of the pharynx and larynx.

3. Injury to the recurrent laryngeal nerve
   The left recurrent laryngeal nerve runs a slightly longer course and tends to be affected more than the right for pathologies. An aneurysm of the aorta can result in compression of the left recurrent laryngeal nerve. In addition, any neck operation will place the recurrent laryngeal nerve at risk, especially if the operative field is close to the tracheoesophageal groove, heart, lungs, or esophagus. In addition, thyroidectomy can put the recurrent laryngeal nerve at risk from damage, but this occurs in approximately 1% of individuals. If one recurrent laryngeal nerve is damaged, it will result in dysphonia (difficulty with speech) and hoarseness. If there is bilateral recurrent laryngeal nerve damage, it can present as a surgical emergency with inspiratory stridor, aphonia, and laryngeal obstruction. It may need to be treated by a tracheostomy in the first instance.

4. Injury to the superior laryngeal nerve
   Injury to the superior laryngeal nerve can occur as a complication of a thyroidectomy. It will result in paralysis of the cricothyroid muscle and anesthesia of the region above the level of the vocal folds. It tends to be, however, the external laryngeal branch that is affected. Therefore, it would affect only the cricothyroid muscle. Some patients may not have any significant consequences of this, while others may have difficulty in changing the pitch of their voice or reduced stamina in their voice. This can have disastrous consequences for those who use their voice in their careers (e.g., singers and public speakers).

5. Vagus nerve stimulation
   Some patients with epilepsy are not able to be seizure free with antiepileptic drugs alone. After trying a variety of different drugs and dosage regimes, it may be necessary to consider vagus nerve stimulation therapy. This entails inserting a small generator, similar to a pacemaker which is surgically inserted to stimulate the vagus nerve 24 h a day.

6. Highly selective vagotomy
   Vagotomy involves cutting of the vagus nerve or parts of it. Previously, it was the gold standard in severe gastric and duodenal ulcer disease. However, with extremely effective triple and quadruple therapy against *Helicobacter pylori*, H2 receptor antagonists, and proton pump inhibitors, the need for this surgery is not as common. Three types of vagotomy can be undertaken—truncal,

selective, and highly selective vagotomy. For a highly selective vagotomy, this only involves denervation of that part of the vagus nerve which supplies the body and fundus of the stomach.

INTERESTING CLINICAL QUESTIONS

Q:
How may an aortic aneurysm or mediastinal tumor affect the recurrent laryngeal nerve, and how will this present in the patient?
A:
Especially on the left-hand side, the left recurrent laryngeal nerve is closely related to the aorta and may therefore affect the nerve with these types of pathologies. Initially, it will present as a cough, as if the mucous membranes of the larynx and trachea were irritated. If the lesion grows, it will then press on or erode the left recurrent laryngeal nerve further. This will lead to hoarseness and potentially paralysis of the vocal folds.

REFERENCE

Yalcxin, B., 2006. Anatomic configurations of the recurrent laryngeal nerve and inferior thyroid artery. Surgery 139, 181–187.

CHAPTER 11

Spinal Accessory Nerve

THE ANATOMY—SUMMARY

The spinal accessory nerve is the eleventh cranial nerve. It is a motor nerve (somatic motor) innervating two muscles—the sternocleidomastoid and trapezius. It has two components—a spinal part and a cranial part. The cranial part of the accessory nerve is from the vagus nerve. However, more recently, it has been shown that not all individuals may have a cranial root (Tubbs et al., 2014). However, when present (in the majority of cases), it joins with the spinal part of the accessory nerve for a short distance. The spinal part of the accessory nerve arises from the first five or six cervical spinal nerves. It has been shown that there is an elongated nucleus which extends from the first seven cervical vertebral levels, which provides the spinal portion of the accessory nerve (Pearson, 1937; Pearson et al., 1964). These branches arise from the lateral side of the spinal cord then form a nerve trunk. This spinal portion then ascends through the foramen magnum passing laterally to join with the cranial root.

As the two nerves join, they then pass through the jugular foramen briefly, along with the glossopharyngeal and vagus nerves. The cranial part then passes to the superior ganglion of the vagus. It then is distributed primarily in the branches of the vagus nerve, specifically the pharyngeal and recurrent laryngeal nerves.

The spinal portion then goes on to supply the sternocleidomastoid and trapezius in the neck.

THE ANATOMY—IN MORE DETAIL

The accessory nerve has two roots—a cranial part and a spinal part. The cranial root arises from the inferior end of the nucleus ambiguus and perhaps also from the dorsal nucleus of the vagus nerve. The fibers of the nucleus ambiguus are connected bilaterally with the corticobulbar tract (motor neurons of the cranial nerves connecting the cerebral cortex with the brainstem). The cranial part leaves the medulla

Clinical Anatomy of the Cranial Nerves. DOI: http://dx.doi.org/10.1016/B978-0-12-800898-0.00011-7

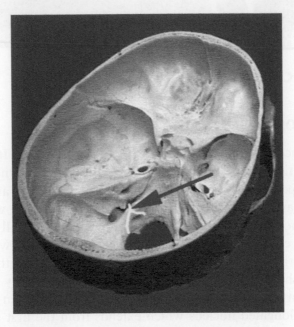

Figure 11.1 The course of the spinal accessory nerve is highlighted by the yellow structure. It shows its two constituent parts—the cranial and spinal portions. The red arrow highlights those.

oblongata as four or five rootlets uniting together, and then to join with the spinal part of the accessory nerve just as it enters the jugular foramen (Figure 11.1). At that point, it can send occasional fibers to the spinal part. It is only united with the spinal part of the accessory nerve for a brief time before uniting with the inferior ganglion of the vagus nerve. These cranial fibers will then pass to the recurrent laryngeal and pharyngeal branches of the vagus nerve, ultimately destined for the muscles of the soft palate (not tensor veli palatini (supplied by the medial pterygoid nerve of the mandibular nerve)).

The spinal root arises from the spinal nucleus found in the ventral grey column extending down to the fifth cervical vertebral level. These fibers then emerge from the spinal cord arising from between the ventral and dorsal roots. It then ascends between the dorsal roots of the spinal nerves entering the cranial cavity through the foramen magnum posterior to the vertebral arteries. It then passes to the jugular foramen, where it may receive some fibers from the cranial root. As it then exits the jugular foramen, it is closely related to the internal jugular vein. It then courses inferiorly passing medial to the styloid process and attached stylohyoid. It is also found medial to the posterior belly

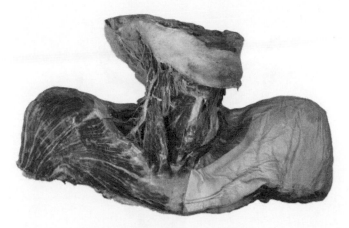

Figure 11.2 An unlabeled image to demonstrate the position of the accessory nerve in relation to other structures.

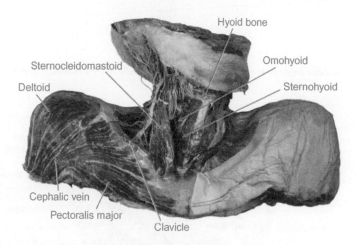

Figure 11.3 A labeled image to demonstrate the position of the accessory nerve in relation to other structures.

of digastric. The spinal root then supplies the sternocleidomastoid muscle on its medial aspect.

The cranial root then enters the posterior triangle on the neck lying on the surface of the levator scapulae at approximately mid-way down the sternocleidomastoid. As it passes inferiorly through the posterior triangle of the neck, and just above the clavicle, it then enters the trapezius muscle on its deep surface at its anterior border. The third and fourth cervical vertebral spinal nerves also supply the trapezius forming a plexus of nerves on its deeper surface (Figures 11.2–11.5).

Figure 11.4 An unlabeled image to demonstrate the position of the accessory nerve in relation to other structures.

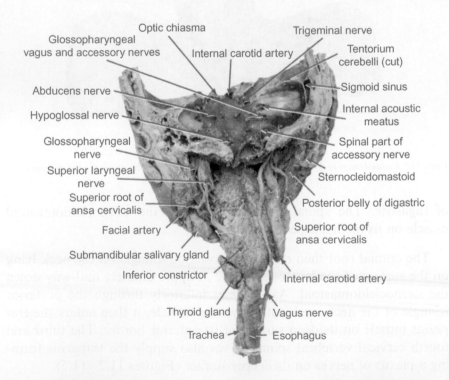

Optic chiasma

Glossopharyngeal
vagus and accessory nerves

Internal carotid artery

Trigeminal nerve

Tentorium
cerebelli (cut)

Abducens nerve

Hypoglossal nerve

Sigmoid sinus

Internal acoustic
meatus

Glossopharyngeal
nerve

Spinal part of
accessory nerve

Superior laryngeal
nerve

Sternocleidomastoid

Superior root of
ansa cervicalis

Posterior belly of digastric

Facial artery

Superior root of
ansa cervicalis

Submandibular salivary gland

Inferior constrictor

Internal carotid artery

Thyroid gland

Vagus nerve

Trachea

Esophagus

Figure 11.5 A labeled image to demonstrate the position of the accessory nerve in relation to other structures.

The Components

The constituents of the spinal accessory nerve have simply been classified as a purely motor nerve. The cranial part of the accessory nerve is classified as a branchial motor nerve, supplying the muscles of the soft palate. The spinal part is classified as somatic motor supplying sternocleidomastoid and trapezius.

However, it has also been shown that the spinal accessory nerve also has sensory fibers from stretch receptors (Echlin and Propper, 1938), non-proprioceptive sensory, and nociceptive fibers (Bremner-Smith et al., 1999).

The Ganglia

There is no known ganglion of the accessory nerve. However, the cranial part of the accessory nerve sends occasional branches to the superior ganglion of the vagus nerve.

The Nuclei

The cranial portion of the accessory nerve arises from the *nucleus ambiguus*, found in the rostral medulla dorsal to the inferior olivatory nucleus. This portion innervates the pharyngeal plexus.

The Important Branches

As previously discussed, the two branches of the accessory nerve are the cranial and spinal branches.

a. Cranial branch—supplies the musculature of the soft palate (apart from tensor veli palatini).
b. Spinal branch—supplies the sternocleidomastoid and trapezius muscles.

THE CLINICAL APPLICATION

Testing at the Bedside

From the clinical perspective, the accessory nerve supplies the *sternocleidomastoid* and *trapezius* muscles, and as such, it is those that are tested when assessing the integrity of the nerve.

The sternocleidomastoid muscle has two functions dependent on whether it is acting on its own or with the opposite side. If the sternocleidomastoid is acting on its own, it tilts the head to that side it contracts and, due to its attachments and orientation, rotates the head so that the face moves in the direction of the opposite side. Therefore, if the left sternocleidomastoid muscle contracts, the face turns to the right-hand side, and vice versa.

If, however, both sternocleidomastoid muscles contract, the neck flexes and the sternum is raised, as in forced inspiration.

The trapezius is an extremely large superficial muscle of the back. It is comprised of three united parts—superior, middle, and inferior. It is involved in two main functions dependent on if the scapula or the spine is stable. If the spinal part is stable, it helps move the scapula, and if the scapula is stable, it helps move the spine. Trapezius is involved in a variety of movements. The upper fibers raise the scapula, the middle fibers pull the scapula medially, and the lower fibers move the medial side of the scapula down. Therefore, trapezius is involved in both elevation and depression of the scapula, dependent on which part is contracting. As well as this, the trapezius also rotates and retracts the scapula.

Testing of the accessory nerve is done as follows:

1. ALWAYS inform the patient of what you will be doing, after introducing yourself and taking a detailed clinical history.
2. When examining a patient, ensure you just observe the patient and try to identify if there is any obvious deformity or asymmetry of the shoulder and neck region. It may be that you will see an obvious weakness or asymmetrical position of the patient's neck and/or upper limbs.
3. First, you can assess the sternocleidomastoid.
4. You can ask the patient to rotate their head to look to the left- and right-hand sides to identify any obvious abnormality.
5. Then, ask the patient to look to one side and test the muscle against resistance.
6. For example, if the patient looks to the right side, place the ball of your hand on their left mandible.
7. Ask the patient to press into your hand.
8. Repeat this on the opposite side.
9. Then, you need to assess the trapezius.
10. First you can ask the patient to raise their shoulders, as in shrugging.
11. Observe any gross abnormality.
12. Then while the patient is raising their shoulders, gently press down on them as they lift their shoulders.

13. Assess any weakness which may be present, noting which side is affected.

Tip!

When assessing the function of the sternocleidomastoid and trapezius, it may help examining the unaffected side first, especially if the patient complains of pain or discomfort on one side. This helps build up trust with the patient but also minimizes causing them any pain or discomfort.

Tip!

Weakness in rotating the head to the left-hand side, when examining sternocleidomastoid, suggests a pathology with the right accessory nerve, and vice versa.

Advanced Testing

The bedside testing should be sufficient in examining the accessory nerve, but the results of this may help direct toward any specialist investigations that may be relevant for the patient. Electromyography studies may be deemed relevant, as well as further cranial and neck investigation by means of CT and/or MRI scanning, dependent on the clinical history and physical examination findings.

PATHOLOGIES

Pathology of the accessory nerve can be divided into various points the nerve leaves the brainstem to its final destination (i.e., the sternocleidomastoid and trapezius).

a. Supranuclear lesion

A supranuclear lesion will result in weakness of both sternocleidomastoid and trapezius muscles, due to bilateral innervation. Within the spinal cord the nuclei can be affected by polio, intraspinal tumors, or amyotrophic lateral sclerosis (motor neuron disease). If the blood supply to the lateral part of the medulla is compromised, for example, due to occlusion of the posterior inferior cerebellar artery, a number of clinical presentations will result. Occlusion of this vessel will affect the trigeminal, glossopharyngeal, vagus, and accessory nerves. Therefore, speech and gag reflex, balance, and facial sensation will also be affected. This is classified as *Wallenberg's syndrome*.

b. Compression at the level of the jugular foramen

Any lesion that exists at the level of the jugular foramen can result in compression of the accessory nerve and also the other cranial nerves entering at that point (i.e., the glossopharyngeal and vagus nerves). This can range from neoplasms to vascular problems (e.g., aneuryms). If the glossopharyngeal, vagus, and accessory nerves are affected together, it is referred to as the jugular foramen syndrome or Vernet's syndrome. Typically, with Vernet's syndrome, there will be loss of taste to the posterior of the tongue, weakness or paralysis of the sternocleidomastoid and trapezius muscles and dysphagia, vocal paralysis and anesthesia of the larynx and pharynx.

c. Lesion of the nerve in the posterior triangle

An isolated accessory nerve lesion is not common. Typically, it is damaged in surgical procedures involving its passageway through the posterior triangle of the neck (e.g., neck dissection or surgical biopsy of tissue).

Torticollis

This condition is a dystonia presenting with an asymmetrical head/neck with the head tilted to one side. A variety of causes exist for it (e.g., congenital, trauma related, infections of the pharynx, base of skull tumors, and cervical vertebral abnormalities). Treatment tends to be alleviating any related pain and gentle exercises to release the stiffness in the neck.

INTERESTING CLINICAL QUESTIONS

Q:

How could you localize where the pathology affecting the accessory nerve is located anatomically?

A:

a. One-sided trapezius muscle weakness + contralateral sternocleidomastoid weakness = upper motor neuron lesion ipsilateral to the affected sternocleidomastoid muscle, above the oculomotor nucleus.

b. One-sided trapezius muscle weakness + unaffected sternocleidomastoid = lesion in brainstem (ventral), cervical spine (lower).

c. One-sided sternocleidomastoid weakness only = brainstem tegmentum or root lesion at the level of the upper cervical vertebrae.

d. One-sided sternocleidomastoid AND trapezius muscle weakness = lesion on the contralateral brainstem, ipsilateral high cervical spinal cord pathology, or issue with the accessory nerve peripherally before it bifurcates to both of these muscles (Manon-Espaillat and Ruff, 1988).

REFERENCES

Bremner-Smith, A.T., Unwin, A.J., Williams, W.W., 1999. Sensory pathways in the spinal accessory nerve. J. Bone Joint Surg. (Br) 81, 226–228.

Echlin, F., Propper, N., 1938. Sensory fibres in the spinal accessory nerve. J. Physiol. 92, 160–166.

Manon-Espaillat, R., Ruff, R.L., 1988. Dissociated weakness of sternocleidomastoid and trapezius muscles with lesions in the CNS. Neurology 38, 796–797.

Pearson, A., 1937. The spinal accessory nerve in human embryos. J. Comp. Neurol. 68, 243–266.

Pearson, A.A., Sauter, R.W., Herrin, G.R., 1964. The accessory nerve and its relation to the upper spinal nerves. Am. J. Anat. 114, 371–391.

Tubbs, R.S., Benninger, B., Loukas, M., Gadol, A.A.C., 2014. Cranial roots of the accessory nerve exist in the majority of adult humans. Clin. Anat. 27, 102–107.

CHAPTER 12

Hypoglossal Nerve

THE ANATOMY—SUMMARY

The hypoglossal nerve is the 12th cranial nerve. It supplies all but one of the intrinsic and extrinsic muscles of the tongue and is a general somatic efferent nerve (somatic motor). It exits the skull at the hypoglossal canal and gives off the meningeal, thyrohyoid, and lingual branches, as well as a component of the ansa cervicalis.

THE ANATOMY—IN MORE DETAIL

The hypoglossal nerve arises from the hypoglossal nucleus. This is found the full length of the medulla, close to the midline. The roots of the hypoglossal nerve itself arise between the olive and the pyramid. These roots then unite to form two bundles which then pass through the hypoglossal canal (in the occipital bone) by piercing the dura mater. It is at this point that the two bundles unite to form the full hypoglossal nerve as a single entity.

After leaving the hypoglossal canal (Figure 12.1), the hypoglossal nerve receives twigs from C1 and C2 containing general somatic motor fibers, as well as some general sensory fibers from the C2 ganglion. It is these fibers that go on further in the neck to supply the strap muscles.

The hypoglossal nerve then continues its journey to the tongue by passing posterior to the internal carotid artery and the glossopharyngeal and vagus nerves. It then passes inferiorly sandwiched between the internal jugular vein and the internal carotid artery. It then loops forward over the occipital artery receiving some fibers from the pharyngeal plexus at that point. As it descends further, it passes over several arteries—the external and internal carotid and lingual arteries. It lies on the hyoglossus inferior to the lingual nerve and submandibular duct then passes inferior to the mylohyoid and digastric. From the hypoglossal nerve are several important branches, which will be discussed later—the meningeal, thyrohyoid, and muscular branches as well as a component to the superior root of the ansa cervicalis (Figures 12.2–12.5).

Clinical Anatomy of the Cranial Nerves. DOI: http://dx.doi.org/10.1016/B978-0-12-800898-0.00012-9

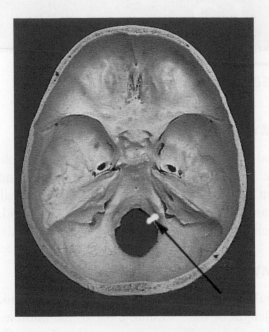

Figure 12.1 The yellow structure, highlighted by the arrow, represents the course of the hypoglossal nerve as it leaves the level of the medulla oblongata to exit the skull at the hypoglossal canal.

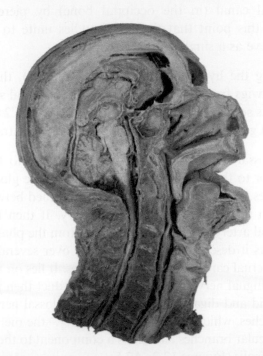

Figure 12.2 An unlabeled image to demonstrate the supply of the hypoglossal nerve in relation to other structures.

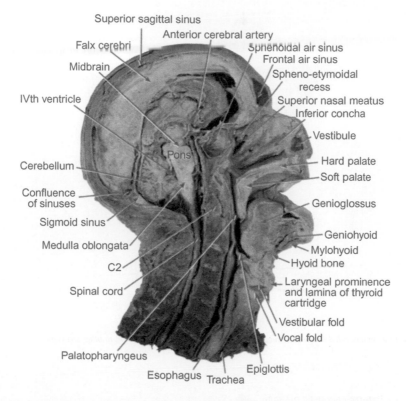

Superior sagittal sinus
Anterior cerebral artery
Falx cerebri
Sphenoidal air sinus
Midbrain
Frontal air sinus
Spheno-etymoidal recess
IVth ventricle
Superior nasal meatus
Inferior concha
Vestibule
Cerebellum
Pons
Hard palate
Soft palate
Confluence of sinuses
Genioglossus
Sigmoid sinus
Geniohyoid
Medulla oblongata
Mylohyoid
C2
Hyoid bone
Spinal cord
Laryngeal prominence and lamina of thyroid cartridge
Vestibular fold
Vocal fold
Palatopharyngeus
Epiglottis
Esophagus Trachea

Figure 12.3 A labeled image to demonstrate the supply of the hypoglossal nerve in relation to other structures.

The Components
Somatic Motor (General Somatic Efferent)
The somatic motor fibers arise from two points—the hypoglossal nerve and C1/2. The bulk of the hypoglossal nerve arises from the hypoglossal nucleus found at the level of the fourth ventricle extending down to the closed part of the medulla oblongata. It is these somatic motor fibers that go to supply the tongue musculature.

General Sensory (General Somatic Sensory)
The general sensory fibers arise from the second cervical spinal nerve. They arise specifically from the C2 ganglion.

The Nuclei
The hypoglossal nucleus is found closely related to the fourth ventricle and extending down to the closed part of the medulla oblongata. The rootlets from this nucleus then gather and exit between the olive and

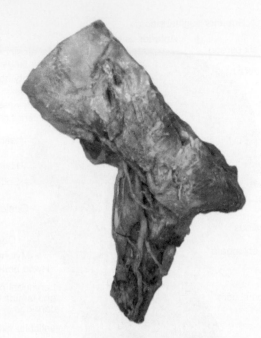

Figure 12.4 An unlabeled image to demonstrate the hypoglossal nerve in relation to other structures.

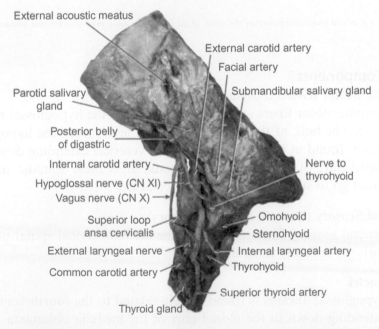

External acoustic meatus

External carotid artery

Facial artery

Parotid salivary gland

Submandibular salivary gland

Posterior belly of digastric

Internal carotid artery

Nerve to thyrohyoid

Hypoglossal nerve (CN XI)

Vagus nerve (CN X)

Superior loop ansa cervicalis

Omohyoid

Sternohyoid

External laryngeal nerve

Internal laryngeal artery

Common carotid artery

Thyrohyoid

Superior thyroid artery

Thyroid gland

Figure 12.5 A labeled image to demonstrate the hypoglossal nerve in relation to other structures.

pyramid. The hypoglossal nerve is divided into those territories of muscles it supplies, i.e., dorsal and ventral laminae.

The hypoglossal nucleus receives information from the precentral gyrus from the opposite cerebral hemisphere via the corticonuclear tract. Connections can also exist with the cerebellum, sensory nucleus of the trigeminal tract as well as the nucleus solitarius.

When the hypoglossal nerve emerges from the brainstem, it is collected into two discrete bundles and acquires its own dural surroundings as it comes to unite to form the single hypoglossal nerve.

The Ganglia
There are no known hypoglossal ganglia. It does, however, receive some general sensory fibers from the ganglion of C2.

The Important Branches
- **Meningeal branches**: These fibers pass upwards through the hypoglossal canal to supply the dura mater within the posterior cranial fossa.
- **Thyrohyoid branch**: This branch arises close to the hyoglossus, passing across the greater cornu of the hyoid bone and supplies the thyrohyoid.
- **Muscular (lingual) branches**: These branches supply the intrinsic muscles of the tongue and also a number of other muscles including styloglossus, genioglossus, geniohyoid, and hyoglossus. However, the fibers to the geniohyoid hitchhike with the hypoglossal nerve, but originate from the first cervical nerve.
- **Superior root of ansa cervicalis**: This branch does not originate from the hypoglossal nucleus, but from the first cervical spinal nerve. It passes to give a nerve to the omoyoid before joining with the second and third cervical spinal nerves, forming a loop called the ansa cervicalis. From this loop, the other strap muscles are supplied by it, namely the sternothyroid, sternohyoid, and the lower belly of omohyoid. The thyrohyoid is innervated by the C1 fibers which also hitchhike with the hypoglossal nerve.

THE CLINICAL APPLICATION
Testing at the Bedside
There are two simple methods which can be used to test the integrity of the hypoglossal nerve, and the tongue it supplies.

1. The first thing to do is simply observe the tongue at rest in the oral cavity. Ask the patient to open their mouth. Observe for any muscle wasting or fasciculations which may be present, as well as any surface lesions. If there is pathology, the tip of the tongue tends to rest to the normal side due to the unopposed pull of the muscle on that side.

2. Then, ask the patient to stick out their tongue. This will assess the muscle function on both sides. If there is a pathology present, the tongue will appear weakened on the affected side and *the tongue will point to the side of pathology*. This is due to the unopposed muscle power of the normal side.

Advanced Testing

Further assessment of the hypoglossal nerve will be directed by the initial presentation, the history taken from the patient, and a clinical examination. From this, it will direct the examining doctor to the relevant investigations which may include, but not limited to, CT, MRI, X-ray, hematological tests, or lumbar puncture. Although idiopathic lesions of the hypoglossal nerve can occur, they are very rare, and in most cases, indicate a space occupying lesion, vascular, neurological, or trauma-related causes (Freedman et al., 2008).

PATHOLOGIES

Pathologies of the hypoglossal nerve can be subdivided based on the level of the pathology.

1. **Supranuclear pathologies**: These types of lesions tend to be highly variable in their presentation, but generally, they produce a very weak and transient weakness of the tongue musculature.

2. **Hypoglossal nucleus pathology**: These lesions tend to be bilateral as the hypoglossal nucleus is located close to the midline, and therefore in close proximity to the one of the opposite side. A variety of presentations can result from a hypoglossal nucleus lesion—weakness and atrophy, complete paralysis, or fasciculations (twitching).

   These can arise for a variety of causes, e.g., due to spinal tumors, infarction, or neurological conditions like syringobulbia, amyotrophic lateral sclerosis, or polio.

3. Hypoglossal nerve pathology: The hypoglossal nerve can be affected anywhere along its length. Tumors tend to be the most

common pathology which affects the hypoglossal nerve (Keane, 1996). These can occur anywhere along its course, but specifically can happen at the jugular foramen. At that site, the accessory and glossopharyngeal nerves could be affected. In addition to this trauma, infection (e.g., basilar meningitis) or neurological conditions (multiple sclerosis, Guillain–Barré neuropathy) are all possible causes which will affect the hypoglossal nerve in its passage to the muscles of the tongue.

INTERESTING CLINICAL QUESTIONS

Q:
What would you notice when examining the tongue of a patient with a longstanding paralysis of their hypoglossal nerve on the right side?
A:
Through time, a paralysis of the hypoglossal nerve, or anywhere along its pathway, from its origin at the hypoglossal nucleus will show a couple of typical features. First, the tongue on the right side will appear wasted and shriveled. Second, when asking the patient to protrude their tongue, it would deviate to the right-hand side. Therefore, the tongue will show these signs ipsilateral to the side of the pathology (if it affects only one side that is!).

Q:
What is the difference between a bulbar and pseudobulbar palsy?
A:
A bulbar palsy is a lower motor neuron lesion affecting the glossopharyngeal, vagus, accessory, and hypoglossal nerves. The pathology

can either be at the nuclear or fascicular level. However, a pseudo-bulbar palsy affects the glossopharyngeal, vagus, accessory, and hypoglossal nerves due to a corticobulbar problem, i.e., upper motor neuron lesion.

Q:
What differences in the appearance and function of the tongue would you notice between a bulbar and pseudobulbar palsy?
A:
With a bulbar palsy, the tongue musculature would be weak and signs of muscle wasting would be evident. Fasciculations would be visible especially when the tongue is at rest in the oral cavity. With a pseudobulbar palsy, the tongue would be paralyzed but with no wasting present initially, nor fasciculations. The speech classically sounds like "Donald Duck." With either of these palsies, a variety of other signs and symptoms are present due to it affecting the glossopharyngeal, vagus, and accessory nerves as well.

Q:
What is a hypoglossal facial nerve transfer, and when may it be used?
A:
If a patient has a facial paralysis that is not amenable to reconnecting the facial nerve, other methods have to be found to reinnervate the face. The nerve which is most frequently used to reinnervate the face for dynamic facial reanimation is the hypoglossal nerve (Ozsoy et al., 2011). It is used because it is extremely close to the main trunk of the facial nerve (accessed surgically as in a parotidectomy), has abundant motor fibers in it, and has been shown to be extremely successful (Conley, 1977; Gavron and Clemis, 1984; Kunihiro et al., 1996; Chan and Byrne, 2011; Ozsoy et al., 2011). Two methods can be used to connect the hypoglossal nerve to the nonfunctioning facial nerve. One approach is to section the whole hypoglossal nerve on one side and transfer it to innervate the distal facial nerve. The other approach which can be used is to divide the hypoglossal nerve and swing part of it to reinnervate the distal facial nerve. This approach can be done with or without a nerve graft connecting these two parts (Conley and Baker, 1979; Dellon, 1992). Reinnervation of the face obtains good results in approximately 70% of patients with a House—Brackmann grade III (Samii and Matthies, 1997).

REFERENCES

Chan, J.Y., Byrne, P.J., 2011. Management of facial paralysis in the 21st century. Facial Plast. Surg. 27, 346–357.

Conley, J., Baker, D., 1979. Hypoglossal-facial nerve anastomosis for reinnervation of the paralysed face. Plast. Reconstr. Surg. 63, 3–72.

Conley, J., 1977. Hypoglossal crossover—122 cases. Trans. Am. Acad. Ophthalmol. Otolaryngol. 84, ORL763–ORL768.

Dellon, 1992. Restoration of facial nerve function: an approach for the twenty-first century. Neurosurg. Q. 2, 199–222.

Freedman, M., Jayasundara, H., Stassen, L.F.A., 2008. Idiopathic isolated unilateral hypoglossal nerve palsy: a diagnosis of exclusion. Oral Surg. Oral Med. Oral Pathol. Oral Radiol. Endod. 106, e22–e26.

Gavron, J.P., Clemis, J.D., 1984. Hypoglossal-facial nerve anastomosis: a review of forty cases caused by facial nerve injuries in the posterior fossa. Laryngoscope 94, 1447–1450.

Keane, J.R., 1996. Twelfth-nerve palsy. Analysis of 100 causes. Arch. Neurol. 53, 561–566.

Kunihiro, T., Kanzaki, J., Yoshihara, S., Saton, Y., Satoh, A., 1996. Hypoglossal-facial anastomosis after acoustic neuroma resection: influence of the time of anastomosis on recovery of facial movement. ORL J. Otorhinolaryngol. Related Species 58, 32–35.

Ozsoy, U., Hizay, A., Demirel, B.M., Ozsoy, O., Bilmen Sarikcioglu, S., Turhan, M., et al., 2011. The hypoglossal-facial nerve repair as a method to improve recovery of motor function after facial nerve injury. Ann. Anat. 193, 304–313.

Samii, M., Matthies, C., 1997. Management of 1000 vestibular schwannomas (acoustic neuromas): the facial nerve preservation and restitution of function. Neurosurgery 40, 684–694.

REFERENCES

Nerve	Point of Entry/ Exit from Brain	Exits Skull	Nuclei	Ganglion	Important Branches	Components	Functions
Olfactory (I)	Forebrain	Cribriform plate of ethmoid bone	No specific nucleus. Olfactory epithelium contain the cell bodies	None	Olfactory epithelium (central processes)	Special sensory	Smell
Optic (II)	Midbrain	Optic canal	Lateral geniculate nucleus	Retinal ganglion cells	Optic nerve; optic tract	Special sensory	Vision
Oculomotor (III)	Midbrain	Superior orbital fissure	Oculomotor nucleus; Edinger—Westphal nucleus	Ciliary ganglion	Motor branches to extraocular muscles; Parasympathetic division	Somatic motor; Visceral motor	Extraocular muscles; Sphincter muscle and ciliary muscle
Trochlear (IV)	Midbrain	Superior orbital fissure	Nucleus of the trochlear nerve	None	None. Only supplies the superior oblique muscle	Somatic motor	Innervates the superior oblique muscle
Trigeminal (V)	Pons	Superior orbital fissure (Va), foramen rotundum (Vb) or foramen ovale (Vc)	Spinal trigeminal nucleus; Pontine trigeminal nucleus; Mesencephalic trigeminal nucleus; Trigeminal motor nucleus	Trigeminal ganglion; Submandibular ganglion	Ophthalmic nerve; Maxillary nerve; Mandibular nerve	General sensory; Branchial motor	Sensation from face, paranasal sinuses, nose and teeth; Muscles of mastication
Abducens (VI)	Pontomedullary junction	Superior orbital fissure	Abducent nerve nucleus	None	None. Only supplies the lateral rectus muscle	Somatic motor	Innervates the lateral rectus muscle
Facial (VII)	Pontomedullary junction	Stylomastoid foramen	Facial motor nucleus; lacrimal nucleus; superior salivatory nucleus;	Geniculate ganglion; Pterygopalatine ganglion;	**Intratemporal** Greater petrosal nerve; nerve to	Branchial motor; Visceral motor;	Muscles of facial expression, stylohyoid,

(Continued)

Nerve	Point of Entry/ Exit from Brain	Exits Skull	Nuclei	Ganglion	Important Branches	Components	Functions
			gustatory nucleus; spinal trigeminal nucleus	Submandibular ganglion	stapedius; chorda tympani **Extratemporal** Temporal; zygomatic; buccal; marginal mandibular; cervical; posterior auricular; posterior belly of digastric branch; stylohyoid branch	Special sensory; General sensory	stapedius, posterior belly of digastric; Parasympathetic innervation of the submandibular and sublingual salivary glands, lacrimal gland and the nasal and palatal glands; Anterior two-thirds of the tongue (and palate); Concha of the auricle
Vestibulocochlear (VIII)	Pontomedullary junction	Internal auditory meatus	Vestibular nucleus; ventral cochlear nucleus; dorsal cochlear nucleus; superior olivatory nucleus	Vestibular ganglion; spiral ganglion	Vestibular nerve; cochlear nerve	Special sensory	Balance for the vestibular component; Hearing for the spiral (cochlear) component
Glossopharyngeal (IX)	Medulla oblongata	Jugular foramen	Nucleus ambiguus; Solitary nucleus; spinal trigeminal nucleus; Inferior salivatory nucleus	Inferior ganglion; Otic ganglion; Superior ganglion; inferior ganglion	Muscular; tympanic; pharyngeal; tonsillar; carotid sinus branch	Branchial motor; Visceral motor; Special sensory; General sensory; Visceral sensory	Stylopharyngeus; Parotid gland for parasympathetic innervation; Taste from the posterior one-third of the tongue; External ear; Pharynx, parotid gland, middle ear, carotid sinus and body

Nerve	Origin	Exit	Nucleus	Ganglion	Branches	Modality	Function/Distribution
Vagus (X)	Medulla oblongata	Jugular foramen	Dorsal nucleus of the vagus nerve; Nucleus ambiguus; Solitary nucleus; Spinal trigeminal nucleus	Superior ganglion; Inferior ganglion	Meningeal branch; auricular branch; pharyngeal branches; superior laryngeal nerve; recurrent laryngeal nerve; carotid branches; cardiac branches; esophageal branches; pulmonary branches; gastric branches; celiac branches; renal branches	Branchial motor; Visceral motor; Special sensory; General sensory; Visceral sensory	Pharyngeal constrictors, laryngeal muscles (intrinsic), palatal muscles, upper two-thirds of esophagus; Heart, trachea and bronchi, gastrointestinal tract; Taste from the palate and the epiglottis; Auricle, external auditory meatus, posterior cranial fossa dura mater; Gastrointestinal tract (to last one-third of the transverse colon), pharynx and larynx, trachea and bronchi, heart
Spinal Accessory (XI)	Medulla oblongata (and spinal cord)	Jugular foramen	Nucleus ambiguous, spinal accessory nucleus	None	Cranial branch; spinal branch	Somatic motor	Innervates the sternocleidomastoid and trapezius muscles
Hypoglossal (XII)	Medulla oblongata	Hypoglossal canal	Hypoglossal nucleus	None. It may however receive general sensory fibers from the ganglion of C2	Meningeal branches; thyrohyoid branches; muscular branches	Somatic motor	Extrinsic and intrinsic muscles of the tongue. Palatoglossus is not supplied by the hypoglossal nerve. It is supplied by the glossopharyngeal nerve

INDEX

Note: Page numbers followed by "*f*" and "*t*" refer to figures and tables, respectively.

A

Abducent nerve, 39, 63, 86
Accommodation, 18, 31–32
Acoustic neuroma, 91–92
Addison's disease, 4–5
Adenocarcinoma, 5*f*, 77*f*
Adrenaline, 29
Afferent, 1, 48, 51–52, 76, 81, 86, 95, 98–99, 110–111
Alzheimer's disease, 4*t*, 5*f*
Ampullary crest, 81–84
Amygdala, 1
Amyloidosis, 77*f*
Anosmia, 4–5
Anterior chamber, 24
Aortic aneurysm, 116
Arachnoid mater, 26
Artery
  celiac trunk (axis), 110
  cerebral, 27, 39, 47, 54*f*, 55*f*
  external carotid artery, 54*f*, 130*f*
  facial, 54*f*, 120*f*, 130*f*
  inferior mesenteric, 110
  inferior thyroid, 106, 108*f*, 112
  internal carotid, 12*f*, 13*f*, 29*f*, 42*f*, 54*f*, 55*f*, 63, 65*f*, 67, 95, 97*f*, 120*f*, 127, 130*f*
  internal laryngeal, 130*f*
  maxillary, 54*f*
  middle cerebral, 54*f*
  subclavian, 106
  superior mesenteric, 110
  superior thyroid, 108*f*, 111, 130*f*
  vertebral, 118–119
Astrocytes, 19
Audiometry tests, 88–89
Auditory cortex, 87
Automated auditory brainstem response (AABR) test, 89
Automated otoacoustic emission (AOAE) test, 88–89
Axons, 7–8, 20, 59–60, 63

B

Basal ganglia, 1, 21
Bell's palsy, 77, 77*f*
Benign postural vertigo, 91
Bitemporal hemianopia, 24
Bone conduction tests, 89
Brainstem, 1, 33, 44–45, 48, 68, 86, 92, 114, 117–118, 123–124
Bulbar palsy, 133–134

C

Caloric testing, 33–34, 92
Canal of Schlemm, 24
Carbamazepine, 59, 102
Cardiac muscle, 112
Catecholamines, 19
Cauda equine, 39
Cavernous sinus, 27, 36*f*, 37, 39, 50–51, 63, 67, 68*t*
Cavernous sinus thrombosis, 36*f*, 67
Celiac plexus, 110, 113
Central nervous system (CNS), 7, 17–26
Cerebellum, 12*f*, 13*f*, 29*f*, 54*f*, 55*f*, 129*f*, 131
Cerebral
aqueduct, 12*f*, 27, 29*f*, 42, 83*f*
  peduncles, 12*f*, 29*f*, 39, 95
Cerebral hemisphere, 83*f*, 131
Cerebrospinal fluid, 7, 9
Cerebrum, 4*t*, 27, 39, 47, 54*f*, 68*t*, 76, 114, 117–118, 131
Cervical, 68*t*, 74, 86, 110, 117–119, 124–125, 127, 131
Cholesteatoma, 91
Cholinergic, 30
Choroid, 10–13, 22, 25
Choroidal epithelial cells, 19
Ciliary
  body, 25
  ganglion, 9, 20, 20*f*, 27, 30, 37–38
  muscle, 15, 27, 30
  nerves, 20, 50–51

Printed and bound by CPI Group (UK) Ltd, Croydon, CR0 4YY

03/10/2024

01040420-0018